AUCKLAND
THAMES
PAEROA
WANGANUI
NELSON
WELLINGTON
WESTLAND
CHRISTCHURCH
DUNEDIN
INVERCARGILL

FARES PLEASE!

Graham Stewart

The Horse, Steam & Cable Trams of New Zealand

Grantham House
New Zealand

For my daughters – Anna & Jane

who shared with me my first experience of riding on a horse-tram, Disneyland, Los Angeles, 1979

By the same author

Tangiwai Disaster, Wilson & Horton, 1972
The End of the Penny Section, A.H. & A.W.Reed 1973,
Collins 1975,
enlarged & revised edition, Grantham House 1993
When Trams were Trumps in New Zealand, 1985
New Zealand Tragedies – On the Track (with Geoff Conly),
1986, 1991, 1992
A Christchurch Album (with W.W. Stewart), 1994
Around Auckland by Tram in the 1950s, 1996
Always a Tram in Sight, 1996
Around Dunedin by Cable Car & Tram in the 1950s, 1997

Photographs with no credit line are from the author's archives and regrettably neither the photographer or the original copyright holder are known.

Published 1997

GRANTHAM HOUSE PUBLISHING
Post Box 17-256, Wellington, 6033
New Zealand

ISBN 1 86934 064 7

Edited by Lorraine Olphert

Printed through Bookprint Consultants Limited, Wellington.

CONTENTS

ACKNOWLEDGMENTS

Many people have assisted in the creation of this book for which I am not only grateful, but delighted for the interest and enthusiasm shown for the history of these vehicles which have long since passed into history.

My father, W.W. Stewart (1898-1976), who made a study of the old steam trams and recorded on film the Takapuna steam tramway; my brother Ian for Auckland research and his continued support; Bruce Gamble and Bruce Pullar of Auckland for photographs; Rodney Wilson, director of the Auckland Museum; Carolyn Stringer, photograph librarian, Auckland City Libraries; Paul Titchener for North Shore archives; the late Jim Housego for photographs; old friend Alan Bellamy of Tauranga for his archival material; Alister Isdale of Thames for his research notes; H.S. Langon of Te Awamutu for photograph; Trevor Terry and Les Turnbull for their Paeroa horse tramway research; John Sullivan and Joan McCracken of the Alexander Turnbull Library; Alan Smith for statistics; Henry Brittain, Richard Gray, Russell Grigg, Ross McQueen and David Todd of Wellington for photographs; Peter Dyer for the use of his steam locomotive drawing; Tom McGavin of Lower Hutt for his continued assistance and support; Mike Johnston of Nelson for his research and knowledge of the Nelson tramway; Anne McEwan and Dorothy Cookson, Nelson Provincial Museum; John Bettle, Graeme Clephane, Trevor Craib, Les Dew and Colin Loach of Christchurch for photographs; Dayle Maffey of Christchurch for his in-depth research; Kerry McCarthy and Jane Barrett, Canterbury Museum; Dawn McMillan, Canterbury Public Library, New Zealand collection; Peter Read, West Coast Historical Museum; Ron Alexander of Rangiora; the late Stan Rockliff who made it possible for me to record the Dunedin cable tramways on camera; Michael Findlay, curator of the Otago Settlers Museum for photographs and assistance; Hocken Library; Bill Campbell of Palmerston for photographs; Cecil Bachop and Ray Hargreaves of Dunedin for photographs; a special thanks to Jim Dangerfield for all his assistance and knowledge; Marilyn Bunce, Keith Douglas and Colin Robertson of Invercargill for their correspondence and knowledge; Lindsay Hazley, photographic curator, Southland Museum & Art Gallery, Invercargill.

Special thanks are acknowledged to four fellow tramway archaeologists for their enthusiasm during the writing of this book for which I record my sincere thanks:

To Gerald Ditchfield of Auckland and Ron Grant of Christchurch who have contributed to this book with definite articles on their specialist subjects – the cable tramways of Dunedin and the New Zealand steam-trams and their origins.

To David Hinman and Bruce Maffey. Together they have researched the large fleet of Christchurch steam and horse-tram trailers once owned by different private companies in the 19th century, many surviving into the second half of the 20th century. During the writing of this book their rekindled interest led them into field trips, unearthing archives long since deposited in attics, visits to research centres, much detective work and long hours of piecing together the known facts.

From overseas, my thanks to Debra Brill of New Jersey, USA, for her assistance with her family company cars, the Brills; John Stevens of New York, who has made a study of the John Stephenson Company of New York; and John Litten, Rail Transport collection, of the National Museum of Science and Industry, London.

To my wife, Anne, for her continued support and encouragement during the researching and writing of this volume – men did stand for ladies on public vehicles in those days!

INTRODUCTION

It is hard to imagine how the lives of city dwellers must have changed when the first horse-trams started plying the streets in the 19th century. Never before had such a smooth and regular form of urban travel – thanks to the steel rail – been possible at such an affordable fare.

This book is about the pioneer tramway vehicles of the 19th century – the horse, steam and cable trams – when most horsepower was generally measured by the number of horses hauling the vehicle. It is a companion volume to my earlier book, *Always a Tram in Sight – the Electric Trams of New Zealand 1900 to 1964*, which told the story of the electric rolling stock. For readers wanting a social history of the horse, steam and cable trams, I recommend them to my earlier book, *The End of the Penny Section* (revised and enlarged edition 1993).

This is the story of the vehicles themselves – their pedigrees where known, as many archival records have been lost, some through fires, others still waiting to be discovered.

They were built by private enterprise, the entrepreneurs of the period. Some were land developers knowing a good tramway service would be vital to selling new suburban residential subdivisions. Many of the companies were under-capitalised and went into liquidation within a few years. The economic depression of the late 1880s and 1890s slowed urban development and saw companies like the Auckland company insolvent. Others struggled with minimum or no dividends for the shareholders. They had been the trail-blazers – building lines through swamps and paddocks to create new suburbs.

The Tramways Act of 1872 gave municipalities the authority, control and legal powers to grant private companies the rights to construct and operate public tramways in the streets. The concessions from local councils were generally for 21 years.

Some of the engineering skills were to lead the world. A young Dunedin engineer by the name of George Smith Duncan designed the "Pull Curve", which enabled cable trams to grip the cable while rounding an uphill curve. Duncan's invention was later adopted by major cable systems in other countries. Duncan also invented an emergency brake known as a "Dolphin" brake which is still used today on San Francisco cable trams. He moved to Melbourne in the 1880s and built their extensive cable-tram system.

Many of the steam-tram locomotives imported into this country, then a far-off colony, had to prove their success to the world on the streets of New Zealand, before they were purchased by the tramway companies in the large centres of the world.

Patents were not strictly registered and controlled in those years. New Zealand companies imported horse-trams from Europe and America, then had local coach builders copy the designs at more economic prices. Many were to stand the heavy loadings of over half a century.

Writing the history of these vehicles has involved much research and the assistance of others, the piecing together of facts from many sources – perhaps similar to Sherlock Holmes solving a crime. Future tramway historians may challenge some of the findings; hopefully this book will stimulate further research and discussion as the years pass.

In December 1862, only eight months after the first passenger tramway line in New Zealand opened between the city of Nelson and its port, the *Otago Daily Times* of 22 December 1862, reported the opening of the Vauxhall Pleasure Gardens in the Dunedin suburb of Vauxhall. The report described the gardens, bush walks, pavilions of pleasure, bars, band rotunda, the concert hall and dance floor, and made a passing reference to a cable car as a means of getting to the gardens from the Vauxhall Wharf which was served by a steam passenger ferry every 15 minutes.

The line was about 150 metres in length on a straight alignment with a steady gradient of about 1 in 3. The motive power is not known, perhaps an arrangement of filling a water ballast tank under the car at the upper station, enabling it to descend while hauling up a counterweight. The gardens closed in 1871 and little is known of this early tramway. Perhaps this was the first cable car in the Southern Hemisphere – who knows – a funicular cliff railway?

A number of horse-trams, steam-tram locomotives and cable trams have been preserved in New Zealand transport museums, thanks to small groups of interested people in the late 1950s and 1960s – otherwise, all would have been lost.

In the centuries ahead the elegant vehicles that have survived will remain a constant reminder of how people once travelled – when travel from a suburban street to a city centre was an excursion.

Graham Stewart

Dates listed under Companies and Corporations are dates the public services started and finished – not the dates when companies and corporations were established or dissolved.

Measurements

As all the trams featured in this book ran in the age of imperial measurements, all the information is given in imperial terms. To convert measurements to metric, use the following table as a guide:

Length	Weight	Currency
1 inch = 25.4mm	1 pound = 0.45kg	£1 = $2.00
1 foot = 305mm	1 ton = 1.016 tonnes	1 shilling = 10 cents
1 chain = 20.177m	**Capacity**	6 pence = 5 cents
1 mile = 1.609km	1 gallon = 4.556 litres	*As at 10 July 1967 when decimal currency was introduced.*

AUCKLAND
Takapuna

Steam Tramway

Takapuna Tramways & Ferry Company Limited

From 22 December 1910 to 26 April 1927

Seven Steam-tram Locomotives, 14 Trailers
Gauge 4-foot 8½ inches
The population of the North Shore of Auckland was 8,000 when the system opened.

The Suburbs and Streets Served by the Steam Trams

From Bayswater Wharf, along Marine Terrace, King Edward Avenue, Lake Road, Hurstmere Road, Kitchener Road, a circular route around Lake Pupuke (via Shakespeare Road, Taharoto Road, Anzac Street to Hall's Corner) and back into Lake Road.

AUCKLAND TO TAKAPUNA.

MONDAY TO SATURDAY.

Steamer Leaves Auckland.	Connects with Tram at Bayswater.
a.m.	a.m.
7.15	*7.30
8.15	*8.30
9.15	*9.30
10.15	h10.30
11.15	h11.30
p.m.	p.m.
12.15	h12.30
1.15	h1.30
2.15	h2.30
3.15	h3.30
4.15	h4.30
5.15	h5.30
d5.45	dh6.0
6.15	*6.30
7.15	h7.30
8.15	h8.30
9.15	h9.30
10.15	h10.30
11.15	h11.30

h Via Hurstmere (Round the Lake)
* Via Northcote Road (Round the Lake)
d Saturdays Excepted

Our parcel rates are very low. Send your luggage through The Takapuna Tramways and Ferry Company, Ltd.

19

TAKAPUNA TO AUCKLAND.

MONDAY TO SATURDAY.

Tram Leaves Milford Rd.	Connects with Steamer at Bayswater
a.m	a.m.
bh6.30	6.55
h7.0	7.30
h8.0	8.30
h9.0	9.30
h10.0	10.30
*10.55	11.30
*11.55	
p.m.	p.m.
*12.55	12.30
*1.55	1.30
*2.55	2.30
*3.55	3.30
*4.55	4.30
*5.55	5.30
h7.0	6.30
*7.55	7.30
*8.55	8.30
*9.55	9.30
a*10.55	10.30
	a11.30

a Does NOT connect with steamer
b Leaves the Depot via Northcote Road at 6.15 a.m
h Via Hurstmere (Round the Lake)
* Via Northcote Road (Round the Lake)

On Public Holidays or special occasions this Time-table is liable to be suspended, when a more frequent service will be run

21

Tramway Sections.

1st Section.—From the starting point at Bayswater to the junction of King Edward Avenue and Victoria Road.

2nd Section.—From such junction to Hauraki Road.

3rd Section.—From Hauraki Road to Anzac Street.

4th Section.—From Anzac Street to East Coast Road.

5th Section.—From East Coast Road to Northcote Road.

6th Section.—From Northcote Road to the Strand.

7th Section.—From the Strand to Hauraki Road.

8th Section.—The same as the Second Section.

9th Section.—The same as the First Section.

43

From the timetable issued by the Takapuna Tramways & Ferry Company Limited, dated 1st January 1917.

Takapuna Tramways & Ferry Company Limited

Steam-Tram

Kerr Stuart Type

Fleet Number: 1 **Name:** *Waitemata*

Year Built: 1909

Into service: 22 December 1910

Withdrawn: 26 April 1927 – scrapped.

Builder: Kerr Stuart & Company Limited, Stoke-on-Trent, England.

Maker's number: 1096

Specifications as built: Wheel arrangement: 2-4-2T. Wheel diameter: 28 inches. Wheel base: fixed: 3 feet 9 inches, total: 13 feet 3 inches. Cylinders: 9½ inches x 15 inches. Stephenson valve gear. Heating surfaces (square feet) – Firebox: 34.5; Tubes: 239.25; total: 273.75. Grate area: 5.42 square feet. Working pressure: 170 lbs. Tanks: 260 gallons. Weight: 15 tons 2 cwt empty; 17 tons 15 cwt full. Fitted with an air-cooled tube condenser on the cab roof. Overhead springs, with coil springs for leading and trailing wheels.

Above: No. 1 being assembled at Bayswater in 1909.

Below: Being used by the contractor for the construction of the line, seen climbing the gradient from Bayswater Wharf.

Above: Takapuna Tramways & Ferry Company Limited Locomotive No. 1 after being fitted with a large wire fender.

Right: Kerr Stuart built No. 1 leaving Bayswater with two trailers in 1919. By now the wheel coverings had been removed. *W. W. Stewart*

Steam-Tram

Kerr Stuart Type

Fleet Number: 2 **Name:** *Bayswater*
Year Built: 1909 – arrived 31 December 1909
Into service: 22 December 1910
Withdrawn: 26 April 1927 – dismantled and boiler sold to a brewery.
Builder: Kerr Stuart & Company Limited, Stoke-on-Trent, England.
Maker's number: 1137
Specifications as built: Wheel arrangement: 2-4-2T. Wheel diameter: 36 inches. Wheel base: fixed: 5 feet; total: 16 feet 6 inches. Cylinders: 12 inches x 16 inches. Walschaert valve gear. Heating surfaces (square feet) – Firebox: 47; Tubes: 383.5; total: 430.5. Grate area: 9.5 square feet. Working pressure: 170 lbs. Tanks: 500 gallons. Weight: 27 tons. Underslung springs on driving wheels, and coil springs on leading and trailing wheels.

Above: No. 2 hauling two four-wheel wagons while working for the contractors during the construction of the line.

Left: Bayswater – No. 2 taking water in Shakespeare Road, Milford. The first trailer behind the steam-tram locomotive is No. 6, which was used briefly after the line closed in 1927 by a Bayswater Point resident as a coal shed, before being purchased by the Wanganui Corporation Tramways and converted to run again as electric tram No. 34 in their fleet.

Left: No. 2 passing O'Neills Cemetery in King Edward Avenue, Bayswater, in 1923. All the wheel skirts had long been removed. *W.W. Stewart*

Steam-Tram

Kerr Stuart Type
Fleet Number: 3
Year Built: 1911 – Left the UK builders on 31 July 1911.
Into service: 11 November 1911
Withdrawn: 26 April 1927 – dismantled, boiler sold to the New Zealand Dairy Company.
Builder: Kerr Stuart & Company Limited, Stoke-on-Trent, England.
Maker's number: 1216
Specifications as built: Wheel arrangement: 2-4-2T. Wheel diameter: 36 inches. Wheel base: fixed: 5 feet; total: 16 feet 10½ inches. Cylinders: 12 inches x 20 inches. Heating surfaces (square feet) – Firebox: 45; Tubes: 385.5; total: 430.5. Grate area: 9.5 square feet. Working pressure: 170 lbs. Tanks: 800 gallons.

Above: No. 3, showing the original iron cab as supplied by the builders Kerr Stuart in 1911. The wheel skirts had been removed.

Above: No. 3 after being fitted with a new iron cab after the First World War.

Above: No. 3 on the last day of service – 26 April 1927.

Below: No. 3 leaving Bayswater Point with three trailers in 1919. Half the cab had temporarily been removed to expose the boiler – giving the steam tram more the appearance of a conventional steam locomotive. *W. W. Stewart*

Steam-Tram

Baldwin Type tram locomotive

Fleet number: 4. The official Baldwin photograph shows the name "*Bayswater*" signwritten on the side waist panel. This would have been for company publicity purposes as No. 2 of the Takapuna fleet had already been given this name.

Year built: 1914

Withdrawn: 26 April 1927 – dismantled, boiler sold to the New Zealand Dairy Company.

Builder: Baldwin Locomotive Works, Philadelphia, USA.

Maker's number: 37168 – 4-18C Type.

Specifications as built: Wheel arrangement: 0-4-0T. Saddle tank. Wheel diameter: 35 inches. Wheel base: 7 feet. Cylinders: 12 inch diameter. 18 inch stroke, with plain side valves. Boiler diameter: 44 inches. Thickness of sheets: 7-1/16. Working pressure: 170 lbs per square inch. Firebox – copper. Length: 56 inches. Width: 34 inches. Depth at front: 45¾ inches. Depth at back: 43¾ inches. Thickness of sheets: ½ inch. Tubes: brass. Number: 134. Diameter: 1½ inches. Length: 6 feet 6 inches. Heating surface – Firebox: 58 square feet. Tubes: 350 square feet; total: 408 square feet. Grate area: 13.2 square feet. Tank: 500 gallons. Total weight in working order: 22½ tons.

This completed the fleet of tram locomotives until after the First World War, during which time they were worked to capacity as it was not possible to order more locomotives from Europe.

Above: The official builder's photograph of No. 4 taken at the Baldwin Locomotive Works, Philadelphia, before the locomotive was shipped to New Zealand. The name "*Bayswater*" would have been placed on the side waist panel for Baldwin publicity purposes only.

Left: No. 4 climbing the 1 in 25 gradient with four trailers c1920. *W. W. Stewart*

Above: No. 4 with the side waist panels rebuilt with tongue and groove timber and the cab windows half covered with wooden panels.

Above: No. 4 with a new iron cab. *W. W. Stewart*

Steam-Trams

Baldwin Type tram locomotives
Fleet Numbers: 5 and 6
Year Built: 1919
Withdrawn: 26 April 1927 – both dismantled, boilers sold to the New Zealand Dairy Company.
Builder: Baldwin Locomotive Works, Philadelphia, USA.
Maker's numbers: 418-C117-52128, 418-C119-52205
Specifications as built: Similar in dimensions to No. 4, except for the steel fireboxes and tubes.

Partly assembled on the Auckland city side on Queens Wharf and taken over to Bayswater by the Auckland Harbour Board's floating crane, the *Mahua.* Took over the bulk of the workload to enable the earlier locomotives to be overhauled which had not been possible throughout the First World War.

Above: No. 6 on the first passenger trip from Bayswater. *W. W. Stewart*

Left: No. 5 shortly after entering service in 1919. *W. W. Stewart*

Left: No. 5 at Milford one Saturday afternoon in the 1920s. *Jim Housego*

Above: No. 6 in Shakespeare Road, Milford, with three trailers. The waist rail of the cab body was much higher than the earlier Baldwin steam- tram locomotives like No. 4, and the curved lower rocker panel was straight.
Left: No. 5 and No. 4 on the last day of service, 26 April 1927. *Jim Housego*

Steam-Tram

Kerr Stuart Type

Fleet Number: 7

Year Built: 1924 – completed 6 May 1924

Withdrawn: 26 April 1927 – sold to Gammon and Company, Sawmillers, Mamaku, who converted the engine into a stationary log hauler.

Builder: Kerr Stuart & Company Limited, Stoke-on-Trent, England.

Maker's number: 4301

Specifications as built: Wheel arrangement: 2-4-2T. Wheel diameter: 34 inches. Wheel base: 34 inches. Cylinders: 12 inches x 20 inches. Stephenson valve motion with balanced Richardson slide valves, and compensated overhead springs throughout. Boiler pressure: 170 lbs per square inch. Total weight: 29 tons.

This locomotive was of orthodox locomotive design, with a cab at one end, bunker, side water tanks, and sand domes on the boiler.

Above: Takapuna Tramways & Ferry Company Limited's No. 7 – the last locomotive commissioned by the company in 1924. *W. W. Stewart*

Above and left: No. 7 taking water at the Bayswater Depot in 1925. *W. W. Stewart*

Above: The last steam locomotive imported into New Zealand for use on a street steam tramway – No. 7 of the Takapuna Tramways & Ferry Company Limited, leaving Bayswater in 1925. *W. W. Stewart*

Steam-Tram Trailers

Fleet Numbers: 1 to 14 (14 cars)
Entered Service:
No's 3 to 8 – 22 December 1910
No's 1 – 15 January 1912, No. 2, and No. 9 – 22 March 1912
No's 10 and 11 – 22 December 1915
No's 12 to 14 – 21 March 1923
The fleet numbers 1 and 2 had originally been designated to the first two steam tram locomotives.
Withdrawn: 26 April 1927
Builder: DSC & Cousins & Cousins, Auckland.
Specifications as built: Double-truck, saloon with smoking partition, saloon windows with the top row drop sash, saloon clerestory roof, tongue-and-groove sides, open fronts. Trailers 12 to 14 were built with glass fronts.

The company had the trailers built so they could be converted to electric trams and equipped with Brush trucks to enable them to be serviced by the Auckland Electric Tramways Company Limited.

Above: Built by the Auckland coach builders DSC & Cousins & Cousins, the original order of six trailers were transported across the Waitemata Harbour in 1910 on a scow to Bayswater. In the photograph (right), the recently launched ferry *Pupuke*, built by George Niccol for the Takapuna Tramways & Ferry Company, can be seen behind the scow.

Trailer No's 1, 2, 4 to 10 – sold to the Wanganui Corporation Tramways – converted to one-man electric trams – series 26 to 34.
The Wanganui Corporation Tramways originally purchased only eight trailers. Then realising one further car was required for their services, purchased trailer No. 6 from a private owner in Takapuna who had intended to convert the body into a summer house. Wanganui also purchased the Brush bogies from the Takapuna Company for use under these trailers when converted to electric trams.
Trailer No's 11 to 14 sold to the Dunedin City Corporation Tramways and converted into electric trams – series 63 to 66. See pages 78, 79, 162 and 163, in the companion volume *Always a Tram in Sight.*

Right: Passengers alighting from two of the original six trailers outside the Mon Desir Hotel in Hurstmere Road, Takapuna, on opening day, 22 December 1910.

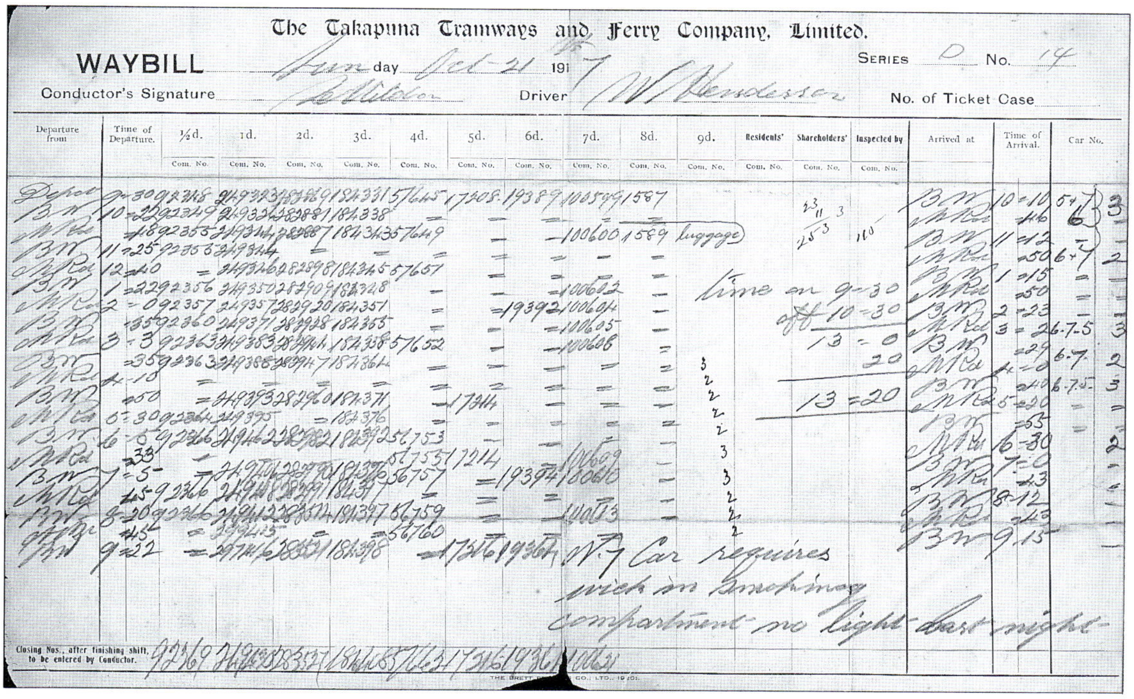
The Takapuna Tramways and Ferry Company, Limited.

WAYBILL Sunday Oct 21 1917 SERIES D No. 14

Conductor's Signature Driver W Henderson No. of Ticket Case

Departure from | Time of Departure | ½d. | 1d. | 2d. | 3d. | 4d. | 5d. | 6d. | 7d. | 8d. | 9d. | Residents' | Shareholders' | Inspected by | Arrived at | Time of Arrival | Car No.

No 7 Car requires wick in smoking compartment no light last night

Closing Nos., after finishing shift, to be entered by Conductor.

Left: A conductor's waybill dated Sunday, 21 October 1917. The driver's name on the waybill is W. Henderson who in later years became mayor of Takapuna. A footnote on the waybill reads: " No. 7 car required wick in smoking compartment, no light last night."

Conductors required were as follows:

- One conductor for one trailer.
- Two conductors for either two or three trailers.
- Three conductors for either four or five trailers.

In 1918 No. 2 had all the seats removed from the non-smoking compartment for the carriage of goods.

Left: Trailers No. 1 in the foreground was one of three trailers built in 1913 as part of the first additions to the fleet.

Below left: Trailer No. 7 (nearest camera), and No. 5 (centre) were part of the original order of six trailers that opened the service in 1910. *W.W. Stewart*

Had electrification proceeded, the company planned to build four new motorised cars. Two car trains would have had one motor car and one trailer; three-car trains two motor cars and one trailer; four-car trains two motor cars and two trailers alternately. It was not proposed to run the cars as single units, but in groups of two, three or four. Five of the present fleet would have been equipped with motors and glass fronts fitted to all the older cars.

Takapuna Tramways & Ferry Company Limited

Right: The plan issued with the prospectus when the company was floated, showing the route of the steam tramway.

Takapuna Tramways & Ferry Company Limited

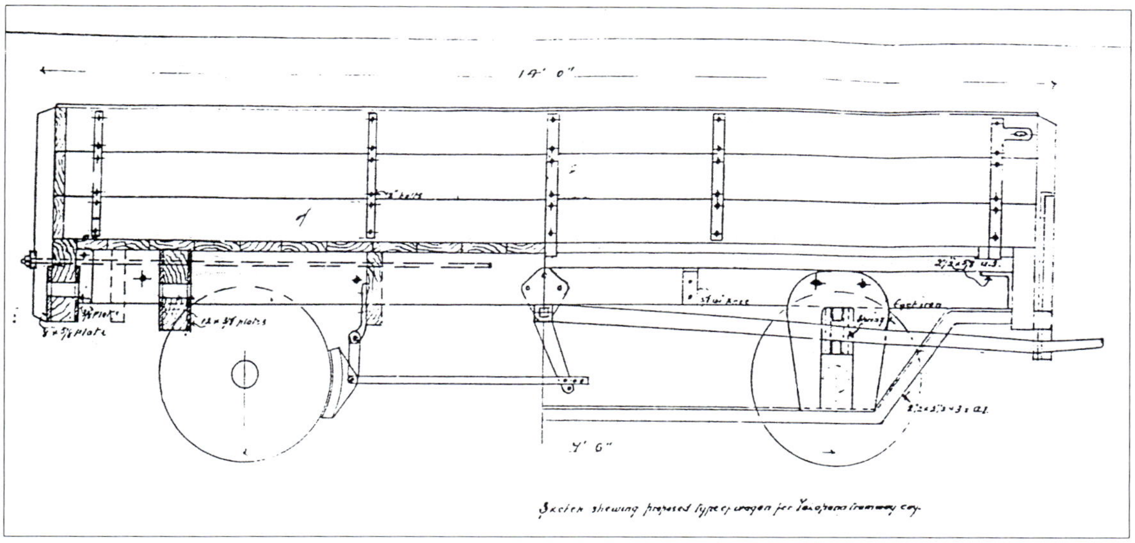

Left: The official drawing of the 14-foot in length ballast wagons built for the contractors and used during the construction of the steam tramway. *Alan Bellamy collection*

Left: No. 1 steam-tram locomotive working with two ballast wagons on the construction of the Bayswater Wharf area in 1909. On the right can be seen the boiler of Kerr Stuart steam-tram locomotive No. 2 and parts in wooden cases ready for assembly. *Price*

Left: The SS *Pupuke* about to berth at Bayswater while a Baldwin steam-tram locomotive shunts two trailers.

Takapuna Tramways & Ferry Company Limited

Right: Steam-tram locomotive No. 5 travelling at speed through Milford c1919, shortly after entering service.

Right: Hall's Corner, Takapuna. A view from the corner of Hurstmere and Lake Roads, looking along Lake Road to Belmont. A steam-tram is approaching the Takapuna shopping centre.

Right: No. 5, the first Baldwin steam-tram built for the Takapuna company, with trailers at the Milford Beach stop, a popular destination for picnic parties in the 1920s.

AUCKLAND

Devonport

Horse Tramway

Devonport & Lake Takapuna Tramway Company Limited

From 25 September 1886 to March / June 1887? 1888?

From the Cash Books of R & R Duder for the month of June 1887, held in the archives of John Webster. The last entry reads:

Friday, 24th June 1887 – Tram horses sent to graze on Browns Island today.

The Devonport Borough Council in May 1887 decided to transfer the Company's interest to Messrs Duder Brothers, who in 1888 secured permission to close down the system for the winter months.

Gauge 4 feet 8½ inches

The population of Devonport was 3,800 when the line opened.

The Suburbs and Streets Served by the Horse Trams

From Devonport ferry jetty (Victoria Wharf), along King Edward Parade, Cheltenham Road, to Cheltenham Beach. Stables and car sheds, corner of William and Lake Streets, Cheltenham.

The single-track horse-tram line to Cheltenham Beach which ran along King Edward Parade and around Torpedo Bay – c1886. *Auckland Institute and Museum*

Devonport & Lake Takapuna Tramway Company Limited

Horse-Trams

Fleet Numbers: 1 and 2 (two cars)
Year Built: 1886
Withdrawn: 1887
Builder: Messrs Cousins & Atkin, Auckland.
Specifications as built: Single-truck, single saloon, curved lower rocker panel, straight upper waist panel, arched roof, open platform at each end.

The *New Zealand Herald* reported the opening of the Devonport Tramway on 11 October 1886:

> On September 25, the portion which has been completed of the Devonport and Lake Takapuna tramway was formally opened for public traffic in the presence of a number of shareholders and guests. The party assembled on board the ferry boat, which left Auckland at 12.30, there being present among others:- Messrs Graves Aicken (chairman of the company), C.W.Cave (secretary, pro tem), D.J.McLeod, W.D.Buddle, Lawson, Hay, Niccol (Mayor of Devonport), Bartley, J.Mays, Trevethick, Knox (engineer to the company), Durance, A.Porter, G.Vosper, W.J.Best, Bell and Glenney. The weather was fine and very favourable for the enjoyment and success of the opening ceremony. On reaching Devonport, the party at once proceeded to the tram line, where the two vehicles of the company were awaiting their arrival, and in one of these, drawn by two excellent horses, the company traversed the whole length of the line to Cheltenham Beach, the distance being between a mile and a mile and a quarter, which was expeditiously gone over, including two short stays at points on the route, which have been arranged to be regular stopping places. At a short distance from the present terminus, at Cheltenham Beach, a siding has been formed to the new stables of the company, a commodious and convenient structure, excellently adapted for the comfort of the horses and the convenience of the traffic. The line is level for nearly the whole distance, the only exception being a short bit, which is at a very easy gradient, and judging from the smoothness and ease with which the cars traversed the distance the road has been thoroughly well laid down. The cars are light and airy vehicles, each being capable of carrying fifty passengers. They were built by Messrs. Cousins and Atkin, from designs by the engineers. The wheels were made at Mr Masefield's establishment, and the whole of the work of the trams was done in Auckland. The cost is about one-third that of imported cars, and though not so elaborate in construction, they are fully strong enough and excellently adapted for their present purpose. The total cost of the plant and equipment was £2300. On reaching Cheltenham Beach the company alighted, and proceeded to inspect the new stables, and afterwards partook of refreshment.

On 29 October 1886 the *New Zealand Herald* reported:

> It is stated that the North Shore tramway traffic is very fair, considering that the weather is still unsettled, and consequently Aucklanders are not tempted to resort to the North Shore for an outing.

The next day the company advertised discount tickets in the *New Zealand Herald* – the news item stated:

> The timetable of the Devonport Tramway will be found in our advertising columns. It will be seen that the trips have been so timed as to suit the arrival and departure of the ferryboats. The company are allowing a discount to purchasers of tickets by the dozen, the prices being 2s 6d for adults; children 1s 6d per dozen; and schools by special arrangement.

Another news item in the same issue drew readers' attention to the fare discounts in the advertising columns of the newspaper and went on to say:

> This liberal action on the part of the company cannot fail to promote tramway traffic, and to induce pleasure-seekers to resort to the North Shore for a cheap day's outing.

Tramways.

DEVONPORT TRAMWAY.

TO AND FROM
VICTORIA WHARF AND CHELTENHAM BEACH
to suit arrival and departure of Ferry Boats.

CHEAP FARES.
FREQUENT SERVICE.
DISCOUNT TICKETS.

Adults, 2s 6d ; Children, 1s 6d per dozen.
Schools by Special Arrangement.

TIME-TABLE FOR NOVEMBER
AND UNTIL FURTHER NOTICE :
From Cheltenham Beach :—A.M. 8.45, 9.28, 10.5, 10.48, 11.33, 11.58 ; P.M. 12.48, 1.58, 2.33, 3.3, 3.28, 4.18, 5.8 5.55.
From Victoria Wharf :—A.M. 9.5, 9 45, 10.15, 11.15, 11.45 ; P.M. 12.15, 1.15, 2.15, 2.45, 3.15, 3.45, 4.45, 5.25, 6.5.

The *New Zealand Herald* 30 October 1886

The *New Zealand Herald* seemed to be doing its best to assist the company. Then on 9 December 1886 the *Herald* reported another reduction in fares:

> In addition to the reduction in fares already made in the case of children, and the fares from wharf to wharf, the Devonport Tramway Company are now reducing the general tramcar fares from 6d to 4d return tickets. The recent reductions in fares have led to an increase in passenger traffic, and the directors have been led to make this further reduction in the confident hope that their endeavours to cheapen the rates of travel will ensure the company increased patronage.

Conveyances.

DEVONPORT TRAMWAY.

TO AND FROM

VICTORIA WHARF AND CHELTENHAM BEACH

to suit arrival and departure of Ferry Boats.

REDUCTION IN FARES.

Adults' Wharf to Wharf Tickets	.. .. ..	1d each
Children's Whole Distance Tickets	.. ..	1d each
Adults ditto	(Return)	4d each

DISCOUNT TICKETS.

Adults, 2s 6d, and 1s 6d per dozen.

Schools by Special Arrangement.

Special Late Cars on requisition.

TIME-TABLE FOR NOVEMBER

AND UNTIL FURTHER NOTICE:

From Cheltenham Beach :—A.M. 8.45, 9.28, 10. 10.48, 11.38, 11.58 ; P.M. 12.48, 1.58, 2.38, 3.3, 3.28, 4.18 5.8. 5.55.

From Victoria Wharf :—A.M. 9.5, 9.45, 10.15, 11.15 11.45 ; P.M. 12.15, 1.15, 2.15, 2.45, 3.15, 3.45, 4.45, 5.25, 6.5.

The *New Zealand Herald* 9 December 1886

*The company had originally planned an extensive network of lines from Victoria Wharf, Devonport, where a three-track terminus had been designed to handle the traffic. The Cheltenham line was to have a passing loop between Church Street and Duders Avenue, and a two-line siding terminus at the beach. The Cheltenham line was to be linked with a line from Victoria Wharf via Victoria Road and Albert Road at the Tainui Road intersection, where the line would continue along Vauxhall Road to the intersection of Wairoa Road at Narrow Neck, with a two-line siding terminus. Another line was to be built along Calliope Road to Stanley Bay, as far as Glen Road where another two-line siding terminus was planned. The longest route was to be to Takapuna, along Lake Road with only one passing loop on the uphill grade before Eversleigh Road, the line terminating at the intersection of Hurstmere Road with The Promenade, where the Mon Desir Hotel now stands. A branch line was to run from Hurstmere Road, down Killarney Street and along Taharoto Road to the intersection of Northcote Road. Two-track sidings were planned for both terminals.

Patronage continued to decline and in an endeavour to save the service, local businessmen Messrs Duder Brothers of Devonport took over the company, but on 19 March 1887 tenders were called for the sale of the company:

> Devonport and Lake Takapuna Tramway Company (Limited)
> Tenders are invited for the purchase of the Property and plant of the above company,viz:-
> Lot 1. Freehold allotment fronting William and Lake Streets, near Cheltenham Beach, with stables and car shed erected thereon.
> Lot 2. The line recently constructed from Victoria Wharf to Cheltenham Beach, of one mile and ten chains, with cars, horses and plant, all new and in perfect working order.
> Particulars to be obtained from Messrs R & R Duder, North Shore, or T & S Morrin & Co. Ltd, Auckland.
> Tenders for the whole or for each lot separately to be sent to the Liquidators care of Adam Porter, Elliot Street, on or before Monday, the 21st inst.
> Adam Porter
> Wm. D. Buddle
> Liquidators.

On 23 March 1887 a meeting of the liquidators of the Devonport and Lake Takapuna Tramway Company took place to consider the tenders received for the company's plant.

When the public service finished is not known, perhaps the cash books of R & R Duder for the month of June 1887 holdthe answer (see page 15).

These cash book entries from the archives of John Webster make interesting reading:

> One broom purchased for 2s 6d, 1 lb of carriage candles for one shilling, the purchase of chaff, bran and oats for the horses. (Three bags of bran cost five shillings in 1887). Recorded on Monday, 20 June 1887 was the return of three bags of oats, 10 bags of chaff, and 1½ bags of bran.

The next day was Queen Victoria's Jubilee – had the service finished?

* Routes described using 1997 street names – from a Plan of Route [*Sic*]), of the Devonport and Lake Takapuna Tramway dated 23 November 1885 held in the author's archives.

Above: King Edward Parade, Devonport – looking towards the corner of Church Street (centre), showing the old Masonic Hotel and the horse-tram line to Cheltenham Beach in the foreground.

*The horses were grazed on Parsons paddock above where the Masonic Tavern is now sited for an annual charge of £50.

*Paul Titchener

Below: The derelict remains of one of the old Devonport and Lake Takapuna Tramway Company's horse-trams which was a landmark on a vacant section at the corner of Eton and Tainui Roads for many years. This photograph was taken in 1903, 16 years after the service closed. Well-known North Shore historian and former mayor of North Shore City, Paul Titchener, told the author that the remains of the tram and those of her sister tram, stood on the section for many years until they both finally rotted away. *Paul Titchener collection*

AUCKLAND

Horse Tramway

The City of Auckland Tramways & Suburban Land Company Limited

Promoters were the St. Heliers & Northcote Land Company Limited

From 11 August 1884 to October 1889

Auckland Tramways

Owners – The Mortgagees, Bank of New Zealand Estates Company
From October 1889 to November 1891
The Bank of New Zealand leased the operation to
Messrs Patterson & Company
From November 1891 to July 1899

The Auckland Electric Tramways Company Limited

From July 1899 to 1902/1903
Phased out as the electric tram services were started

Gauge 4 feet 8½ inches
The population of Auckland was 32,000 when the system opened.

The Suburbs and Streets Served by the Horse Trams

From the bottom of Queen Street, up Queen Street (up line only), Wellesley Street West (single track only), Hobson Street (single track only – temporary loops at Wellington Street and top of Pitt Street), Pitt Street, Karangahape Road to Ponsonby Road 11 August 1884.
Wellesley Street West and Hobson Street – double track opened 5 January 1886.
Pitt Street – double track opened 13 December 1885.
Freemans Bay branch – Customs Street West to Fanshawe Street 14 November 1884. To Patteson Street 14 November 1884. (originally planned to be part of a western circle line from Three Lamps, down College Hill)
Along Ponsonby Road to the junction of Jervois Road and College Hill at the Three Lamps, down Jervois Road to Wallace Street 9 February 1885.
Along Karangahape Road from Pitt Street and up Symonds Street to Khyber Pass Road 7 November 1885.
Wellesley Street East, Symonds Street to Karangahape Road – double track 21 December 1885.
Khyber Pass Road to Grafton Road 21 December 1885.
From Grafton Road, down Khyber Pass Road into Broadway, Newmarket, to a siding in Remuera Road 27 February 1886.
Choral Hall branch – from the top of Wellesley Street East, down lower Symonds Street (now Anzac Avenue), to Grafton Road 17 March 1886.
Manukau Road – from Remuera Road to Greenlane Road 29 February 1888.
Onehunga – Greenlane Road to the Onehunga Wharf 10 March 1903. The horse-trams used the electric tram rails for a temporary service until the electric trams started running.

Depots and Stables:

Customs Street West – the Kamo Coal Company hired to the Tramway Company a portion of their depot for the stabling of tram horses until the Wallace Street depot was completed in August 1884.
Ponsonby – Corner of Jervois Road and Wallace Street. Horses: 28 January 1885. Trams: 9 February 1885.
Hobson Street – Twelve-horse stable for leader horses – March 1885.
Epsom – Potters Paddock, corner Manukau and Greenlane Roads 29 February 1888.

City of Auckland Tramways & Suburban Land Company Limited

Horse-Trams

Fleet Numbers: Probable numbers: 1 to 11. Bingham 1898 report listed 11 closed cars to carry 24. Known numbers 2, 3, 5, 11.
Year Built: 1883
Withdrawn: c1902
Builder: John Stephenson Company Limited, New York, USA.
Specifications as built: Single-truck, single saloon with longitudinal seats, curved lower rocker panel, straight upper waist panel, clerestory roof, open fronts.Small five-window saloons, two horses.

Above from left: No. 11 rounding the corner into Wellesley Street East. *Middle:* Like the photograph of No. 11, heavy roof advertising boards appeared within a few months of the service opening. *Above:* One of the small five-window saloons before roof advertising. In each of these photographs, the original sign-writing was still visible. On the letter board, "Newton & Ponsonby", and on the waist panel, "City of Auckland Trar..ways".

Above: This well-known classic photograph taken by pioneer Auckland photographer W.B.Beattie Senior (the author worked with his son, W.B. Beattie Junior, on the *Weekly News* in the 1950s) of a five-window single-saloon car climbing the Wellesley Street West gradient to Hobson Street in January 1900 – shows one of these pioneer Auckland horse-trams still robust after 16 years of working the inclines of the Wellesley Streets out of Queen Street. *W.B. Beattie Senior*

**New Zealand Herald* 6 February 1884:
Four of the tramway carriages for the Auckland tramways have just been landed from the *Opawa* and 24, in adddition, have arrived in Wellington and will be landed here in less than a month... being fitted up at Messrs Morrin & Co's warehouse, High Street.

Right: Still sitting on its wheels fifty years after retirement – one of the five-window saloons in a sad state of decay on Herne Bay point in the early 1950s. *Graham Stewart*

Horse-Trams

Fleet Numbers: Probable numbers: 12 to 21, 25 to 31. W. Gentry Bingham 1898 report listed 16 closed cars to carry 36 passengers. Known numbers 15, 16, 17, 18, 31.
Year Built: 1883
Withdrawn: c1902
Builder: John Stephenson Company Limited, New York, USA.
Specifications as built: Single-truck, single saloon with smoking and non-smoking compartments, longitudinal seats, curved lower rocker panel, straight upper waist panel, clerestory roof, open fronts. Large eight-window saloons – used with two and three horses, dependent on passenger loadings.

Left: At the Three Lamps terminus, Ponsonby, the conductor at the rear is J. Glynn and the driver, second from right, C. Nankervis. The horse nearest the camera was named "Old Melody".

*Bingham 1898 Report listed 27 saloon cars. *New Zealand Herald* 6 February 1884 stated a total of 28 saloon cars were imported from John Stephenson, New York,

These photographs of the large eight-window single-saloon cars working on the western and eastern lines of Auckland illustrate how the heavy advertising boards were added, regardless of the extra weight, to each horse-tram. Like today, advertising revenue was important.

Left: The last survivor was sadly not saved for preservation. An eight-window single-saloon car at the Tramway Workshops, Epsom, thirty years after the horse-trams were retired. *New Zealand Herald*

All saloon cars had Stephenson patented cushioned with rubber running gear.

Horse-Trams

Fleet Numbers: Probable numbers: 21 to 23, 32 to 34. W. Gentry Bingham report 1898 listed 6 open cars to carry 36 passengers. Known numbers 21, 22 – arched roof; 32, 33, 34 – clerestory roof.
Year Built: c1880s
Withdrawn: 1902
Builder: Unknown. Some by Cousins & Atkin, Auckland.

Specifications as built: Single-truck, open toast-rack with cross-bench seats, entry-exit running boards, glass end bulkheads, canvas weather blinds, open fronts. Five and six bench.

The Bingham report included an inventory of the Auckland Tramways Company as at 31 March 1898 and listed a total of 33 horse-trams. As there are two photographs of horse-tram No. 34, one car must have been either written-off by this date or one fleet number had not been allocated.

The *New Zealand Herald* reported on 24 October 1885:

> The first summer car placed in service yesterday, weighs 34 cwt, licensed to carry 30 passengers. A second being built by Cousins & Atkin.

Open summer two and three-horse, six cross-bench seat horse-trams all photographed at the Three Lamps terminus in the 1880s. As the lower photograph shows, roof advertising boards soon added weight to the load of the horses.

New Zealand Herald 9 April 1884:

> The new tramway stables for the Auckland Tramway Company, to be erected at the corner of Wallis [*sic*] Street and Jervois Road, will be constructed of brick or concrete, two storeys in height, with corrugated roof. They will be 130 feet in length by a breadth of 79 feet, giving accommodation for 70 horses. On the upper storey will be situated the hay loft and stores for feed, also eight feed shoots. There are four large entrances or exits with sliding doors, also two with swing doors. The carriage shed will be 110 feet in length by 44 feet in breadth, one storey in height, having concrete foundations, and being built of corrugated iron walls and roof.

Left: This horse-tram, according to the notice on the front apron, was licensed to carry 30 passengers. A light-weight, six cross-bench open horse-tram. A stable-hand holds the two-horse team while the photographer recorded the occasion at the Ponsonby depot.

Above: Another open cross-bench horse-tram of a slightly different design with an arched roof at the Three Lamps Ponsonby terminus. The conductor, second from left, complete with cash bag, watch chain and whistle.

Left: No. 34 was a clerestory-roof open car with six cross-bench seats. This is the highest fleet number Auckland horse-tram recorded by photographers of the period.

Above: Other designs used in Auckland included two-compartment Palace-style cars with entry/exit from either the side running-board or from the end platforms.

Right: No. 24 was another varation of the six-bench open car with a clerestory roof. Entry/exit was from both the side running-boards and the end platforms.

Below: The roll-over-back cross-bench seats on this six-bench open horse-tram show clearly in this downtown view. The summer horse-tram was waiting at the foot of Queen Street for passengers arriving from the North Shore on the paddle-steamer ferry *Britannia*, seen berthing in the background. *Alexander Turnbull Library 33704½*

Above: No. 18, an eight-window saloon horse-tram at the foot of Queen Street in the mid-1880s. The conductor is checking his tickets on the back platform and the horses are resting in readiness for the journey to Newmarket. Wingates Hardware was still in the same building on the right when the electric trams were withdrawn in 1956.

New Zealand Herald 31 May 1884:EXPERIMENTS WITH BRAKES. Yesterday afternoon one of the American tram cars, which has been fitted up, was taken from the depot along Albert Street by a pair of horses, and placed on the Hobson Street and Wellesley Street section of the tramway. The object was to test on the steepest grade of the line – that of Wellesley Street – an adaptation of the slipper brake, designed by Messrs Stewart and Bunter, the tramway engineers, with which it is proposed to furnish each of the cars in addition to the ordinary brake, so that in the event of an accident to the one the other would be a stand-by, and also be available for the ordinary and steep grades of the line... The car was run up and down the Hobson Street section, with two horses in, at a good rate... The car was then tried on the Wellesley Street down grade without the horses, the brakes acting very effectively...

Below: Looking up Queen Street – a three-horse summer toast-rack horse-tram crossing the intersection of Customs and Queen Streets in 1899. Dilworth Building now stands on the right. *Beattie & Saunderson*

Right: Looking down on the intersection of Queen Sreet and Customs Street West with the port beyond in the 1890s. The double-track horse-tram line in Queen Street reverts to single track just before the intersection and a single track curves round into Customs Street West. The main line continues down to Queens Wharf where two horse-trams can be seen standing at the end of the line. *Burton Brothers, Alexander Turnbull Library F43083½*

The *New Zealand Herald*, 14 November 1884:

> The authority to allow the tramcars to run to Freemans Bay appears to have been at last granted by the Government. On and after today the cars will run from Billington's corner, Freemans Bay, to the post office at Newton from 7.10 am to 10.50 pm. The first car will leave Newton P.O. at 7.40 am and the last reaches Freemans Bay at 11.20 pm. On Saturday additional cars will run.

The *New Zealand Herald* reported on 27 November 1884:

> The Tramway Company are now carrying over 6,000 passengers per week. It is found that the extension along Customs Street West to Billington's store, Patteson Street, Freemans Bay, has led to an increase in traffic ...

Above: A busy 1890s in Queen Street; Fort Street is on the left. A schoolboy and a youngster were hitching a free ride on the back steps of a Paterson horse-bus. In the centre distance, No. 34, an open summer horse-tram, is moving up Queen Street, while on the right an opposition to the trams a Newton and Ponsonby horse-bus is heading downtown.

Above: Looking up Queen Street from the Customs Street intersection in the late 1880s. *Courtesy E.A. Wildy, Auckland Institute and Museum*

Below: The same view in the late 1890s. An inward bound horse-tram in the foreground is moving from the down double-track section to the single line which ran across the intersection before reaching the double-track section in lower Queen Street. *Auckland Weekly News*

Tramways.

CITY OF AUCKLAND TRAMWAYS, AND SUBURBAN LAND CO (LIMITED).

NOTICE.

On and after MONDAY, the 18th instant, the following alterations will be made in the Time-table of the Ponsonby Circuit (only).

Cars will leave Lamps and Queen-street respectively every half-hour between the hours of 7 a.m. and 8 a.m.; every quarter of an hour between the hours of 8 a.m. and 10.30 a.m.; every ten minutes between the hours of 10.30 a.m. and 7 p.m.; with the following extra Cars, viz., 5.5, 5.15, 5.35, 6.15.

For particulars of running after above, see Time-table, to be had from all Conductors, and at the Company's Office gratis.

There will be no alteration in the running of the Queen-street, Newmarket, Onehunga, and St. Helier's Cars and 'Buses.

For College Hill 'Buses see Company's Time-table.

On and after MONDAY, the 18th instant, 1d extra will be charged for Transferring Passengers from Ponsonby to Newmarket or vice versa.

October 15, 1886.

The *New Zealand Herald* 15 October 1886

Right: In Queen Street a two-horse five-window saloon horse-tram is passing Swanson Street on the right as a Landau carriage moves off the tram tracks in the foreground.

Below: Queen Street in the late 1890s. The small two-horse tram was heading for Newmarket, the three-horse larger tram to Ponsonby. Wyndham Street is on the left.

Right: Summer horse-tram No. 33 passing Wyndham Street on a journey up Queen Street surrounded by horse-buses. *Auckland Public Library 220*

Looking down Queen Street from the Victoria Street intersection.

Above: c1885 – a two-horse five-window saloon horse-tram is on the right. *Alexander Turnbull Library F65501½*

Below and right: c1890s – the large roof advertising boards were far from elegant.

Right: Looking up Wellesley Street West from the intersection of Queen Street shortly after the horse-tram service to Ponsonby started in August 1884. The single-line up Wellesley Street West was duplicated in January 1886. The line in the foreground curving round into Wellesley Street East is covered with road metal. The eastern line did not open for traffic until December 1885. *Auckland Public Library 779*

The original proposal was for an underground cable tramway system for Auckland, which was rejected because of the high cost of installation. The cost of a cable line for College Hill alone would have cost £10,000 and was still being considered when the horse-drawn trams started.

Right: Not a horse-tram in sight – the final horse-tram track arrangement in Wellesley Street East (foreground) and West. Single curved tracks ran from Queen Street, then went into double tracks in the Wellesley Streets. The *NZ Gazette* of January 1887 listed an extension "from the foot of Wellesley Street West and East in a southerly direction to and as far as the foot of the hill in Upper Queen Street (17½ chains)." There is no record of this line ever being constructed. *Burton Brothers, Museum of New Zealand Te Papa Tongarewa C11201*

Above: The Wellesley Street West gradient in 1887. The Wellesley Billiard Rooms, as reported in the *New Zealand Herald* story below, are on the right below J. Watkinson, the tailor.

The *New Zealand Herald* reported on 28 August 1884:

> At ten minutes past five yesterday afternoon the tramcar ascending the Wellesley Street East [*sic*] hill came to a full stop opposite the Wellesley billiard-room, and commenced rolling backwards down the declivity, dragging with it the three horses. Some of the passengers and the conductor jumped off and blocked the wheels with pieces of road metal, and by dint of considerable persuasion, again started the vehicle. The horses bravely struggled up the hill, encouraged by the "git ups" of the postilion, and bid fair to accomplish the rest of the journey with success. When, however, in their endeavour to negotiate the pinch opposite Dr. Dawson's house they again came to a standstill, and the music recurred *da capo.*

The *New Zealand Herald* of 1 September 1884:

> An accident of a serious nature occurred in connection with the tramway on Saturday evening at seven o'clock to a boy named Breen, employed to ride the leading horse, which is hooked on at the bottom of the Wellesley Street East [*sic*] hill, to assist in dragging the car to the high ground on Hobson Street. It would appear that the chain which it is usual to attach to the pole in some manner gave way, causing the horse to swerve and throw its rider heavily to the ground. The poor lad was picked up insensible, with a very severe cut extending across the lower lip, apparently having been effected by the horse's hoof, and was carried over to Dr. Tennent's. His injuries having been attended to, he was sent to his home in Chapel Street in a cab...

A few days later on 2 September 1884, the *New Zealand Herald* reported further problems on the Wellesley Street West gradient:

> The Wellesley Street hill would appear to be the rift in the lute of the Tramway Company. Yesterday afternoon, when the crowded car essayed the ascent, repeated mishaps not included in the programme, occurred to vary the monotony of the journey. Between five and six o'clock the chestnut leader jibbed entirely, and manifested signs of being "full up" of the hill, refused to continue the contract, and had to be led away. Borrowing a metaphor from our banking institutions, there being evidence of an overdraft, the driver had recourse to the brake, and put on the screw, which prevented the vehicle from running back to Queen Street. Eventually, a fresh start having been effected, several passengers assisted the energetic Secretary, and lent a helping hand to push the tram-car along.

Above: Safely around the corner in Hobson Street is summer horse-tram No. 34, the last of the fleet. Wellesley Street West is on the left.

Above: Apart from the Baker's cart turning into Lorne Street, all the horse-drawn vehicles are public trams and buses. The Ponsonby horse-tram in the foreground has a leader-horse with leader boy astride starting to take the Wellesley Street West grade. In the distance, passing the library, a Newmarket tram is also being assisted with a leader-horse.

Left: An inbound eight-window saloon horse-tram from Newmarket passing down Wellelsey Street East during the visit of the Duke and Duchess of Cornwall & York in 1901.
Above: Wellesley Street West on a summer's day in 1897.

Right: In the distance No. 15 starts the long haul up Wellesley Street East, while in the foreground a two-horse five-window saloon bound for Ponsonby rounds the corner from Queen Street to Wellesley Street West. *Otago Settlers Museum*

The *New Zealand Herald's* report of a horse-tram that ran away down Wellesley Street East in August 1902:

> An occurrence of an exciting nature took place in Wellesley Street shortly after seven o'clock last evening, which resulted in the wrecking of a tramcar, and in its conductor, a young man named John Davis, receiving injuries of a nature which subsequently necessitated the amputation of his right foot. It appears that as a three-horse tramcar running on the eastern circuit, and driven by Michael Hogan, was coming down Wellesley Street East the brakes failed to grip the wheels sufficiently. The driver at first thought that some one was tampering with the rear brakes, but this impression the conductor found to be wrong. By this time the speed of the car had been accelerated greatly, and during its mad career to the bottom of the hill, both conductor and driver displayed great presence of mind, Davis throwing out sacks and gear on to the track in the hope of stopping the car, but their combined efforts were useless. On reaching the Queen Street level the car jumped from the track and landed with a deafening crash against one of the steel centre poles recently erected in the middle of the street in connection with the electric tramway construction. The pole crashed between the horses, breaking the hind legs of one of them, and it had subsequently to be destroyed.
>
> The force of the impact was terrific, and the upright pole penetrated the car for a couple of feet, pinning Davis to the front wall. When the man was rescued from his perilous position it was seen that his right foot was crushed to a pulp, and that it was hanging to the ankle by a few shreds...
>
> The driver escaped practically unhurt, beyond a bruised knee and slight injury to his toes.
>
> Fortunately there were but few passengers in the car, and these jumped out and escaped injury, with the exception of one unknown man, who got a severe bruising, and slightly hurt his knee ...

A contemporary newspaper artist's impression of the runaway horse-tram hitting the steel pole at the intersection of Queen and Wellesley Streets in August 1902.

The extension to Epsom was single track and ran on the left-hand side of the road with seven loops. From Newmarket, the loops were as follows; just past Mortimer Pass, just before the Great South Road intersection, just past Owens Road, at Mount St John Avenue, at Ranfurly Road, just past Market Road, and just before the four track car shed and stables at Potters Paddock (now Alexandra Park).

Left: Looking along Khyber Pass Road from the top of Symonds Street. The horse-tram track and the loop in Khyber Pass Road were used for the service to Newmarket. The parade was for the visit of Imperial and Indian troops. The Imperial troops had arrived on the *Britannic* and the Indian troops on the *Dalhousie* in February 1901.

Right: No. 22, a cross-bench open horse-tram and a single-saloon behind on the left, waiting in Manukau Road at the entrance to Potter's Paddock (Alexandra Park) in July 1898, for the finish of the annual Auckland v Wellington rugby match. It was a very wet day as the muddy road shows. *Auckland Weekly News*

Above: Digging up the old horse-tram tracks for the laying of the electric tracks in 1901. In Queen Street by Fort Street, a pick and shovel gang at work.

Above: In Wellesley Street East the old tracks are being lifted while new tracks for the electrics are being laid at the Queen Street intersection, and horse-trams continue to operate to Ponsonby from the foot of Wellesley Street West.

Left: Horse-trams resumed running on the electric tracks once the new tracks had been laid until the electric service started in November 1902.

Right: Horse-trams standing in Jervois Road, Herne Bay, while the old Ponsonby depot on the corner of Wallace Street was being rebuilt for the electric trams in 1902. Brand new electric double-decker No. 17 looks smart and modern beside the veteran horse-cars.

THAMES

Steam Tramway

The Grahamstown & Tararu Tramway Company

2 December 1871 to November 1874

Gauge 3 feet 6 inches

The population of Thames was 12,238 when the line opened.

The Suburbs and Streets Served by the Steam Trams

From Curtis' Wharf, in Albert Street, Grahamstown, around the foreshore to Tararu Point and wharf.

In August 1871, the Superintendent of the Auckland Province granted for a period of 10 years to the Grahamstown and Tararu Tramway Company, the right to construct, maintain and use a tramway to be laid down on the highway between Grahamstown and Tararu, the said tramway to revert to the Superintendent absolutely at the end of 10 years.

The line was laid down by J.S. Macfarlane and Company of Auckland for around £10,000 for line and rolling stock.

After closure, the line slowly deteriorated as ownership was challenged by the Thames Borough Council after the assets had been taken over by the Bank of New Zealand. The line was not removed until June 1885.

The coastal route of the Grahamstown and Tararu tramway which ran from Albert Street, Grahamstown, to Tararu Point. *H.S.Langon collection*

Grahamstown & Tararu Tramway Company

Steam-Tram

Chaplin Type
Year Built: 1870
Withdrawn: 1874
Builder: Alexander Chaplin and Company, Cranstonhill Engine Works, Glasgow, United Kingdom.
Believed to be Maker's No. 1182. Outshopped in April 1870 to the order of G.E. Stevenson of Liverpool for export.*
Imported by the Bay of Islands Coal Company, Kawakawa, placed in service on 28 January 1871. The first steam locomotive to work in the North Island. Found to be not powerful enough for the coal company requirements and sold to the Grahamstown and Tararu Tramway Company. Shipped to Auckland and altered to 3-feet 6-inch gauge at the Fraser and Tinné Phoenix Foundry.
Specifications as built: Wheel arrangement: 0-4-0. Gauge: 4 feet 8½ inches. 12 horsepower. Cylinders: 6 inches x 13 inches. Geared 1 to 2. Driving wheels diameter: 3 feet 1 inch. Wheelbase: 6 feet 4 inches. Overall length: 13 feet 3 inches. Weight: 8 tons. Vertical water tube boiler steamed at 85 to 100 lbs to the square inch pressure. The two cylinders were mounted vertically ahead of the boiler and drove a crank shaft which was geared to one of the axles. The wheels were coupled by circular section side rods.

* Ron Grant research.

The vertical-boiler Chaplin locomotive working on the Bay of Islands Coal Company line at Kawakawa before it was sold to the Grahamstown and Tararu Tramway Company. *W.W. Stewart collection*

Given the nickname *Jumping Jenny* by the locals because of the unevenness of the coastal track.
The *Thames Star* reported on 11 November 1874:

> The Tararu Tramway closed yesterday on account of there not being sufficient traffic to make it pay now the daily passenger trade to Tararu has ceased through the sale of the *Golden Crown* steamer. The steamers at present in the trade are generally able to bring passengers to Grahamstown and Shortland on account of their light draught.

The *Thames Advertiser* on 26 April 1879 reported the following:

> The little engine which formerly did service on the Grahamstown and Tararu Tramway, and which had been purchased by Messrs O'Brien and Company, the railway reclamation contractors, was taken out of the shed where it has reposed since the tramway ceased operations, cleaned, and steam got up. The diminutive iron horse worked very smoothly, and its performance satisfied its new owners, who intend employing it to draw the trucks which are to be used to convey the materials from the Piako hill to the beach.

The newspaper report did not identify the engine as either the Chaplin or the Fraser and Tinné.

Below: *New Zealand Herald* August 1871

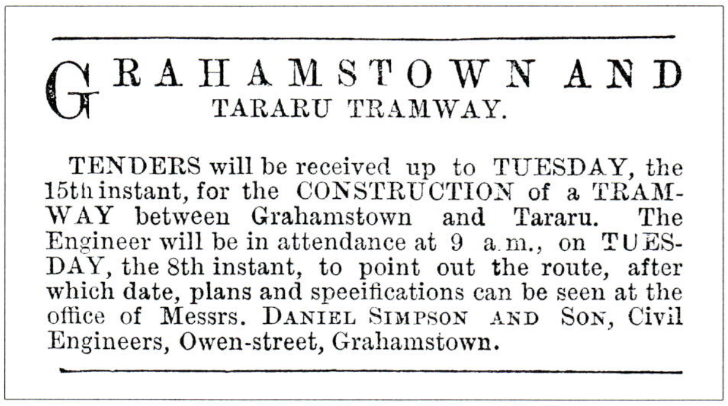

GRAHAMSTOWN AND TARARU TRAMWAY.

TENDERS will be received up to TUESDAY, the 15th instant, for the CONSTRUCTION of a TRAMWAY between Grahamstown and Tararu. The Engineer will be in attendance at 9 a.m., on TUESDAY, the 8th instant, to point out the route, after which date, plans and specifications can be seen at the office of Messrs. DANIEL SIMPSON AND SON, Civil Engineers, Owen-street, Grahamstown.

The *Thames Mail* reported on 11 September 1871: About 60 men started work this morning on the tramway between Grahamstown and Tararu.

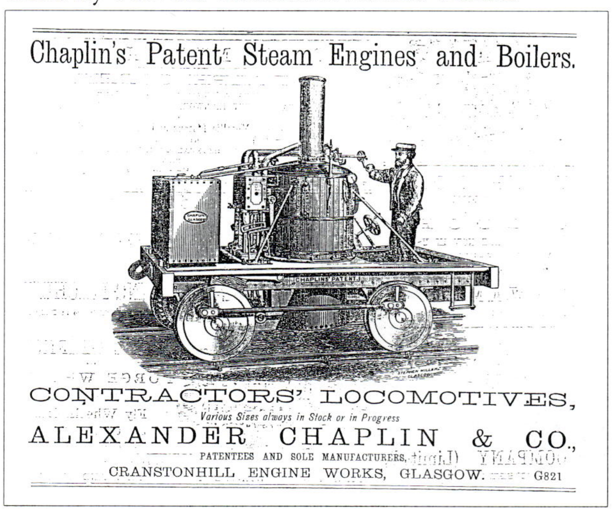

Right: Advertisement in the *Engineer* magazine in 1875. *Ron Grant collection*

Steam-Tram

Fraser and Tinné Type
Year Built: 1872
Withdrawn: 1874
Builder: Fraser and Tinné Foundry, Auckland.
Specifications as built: Wheel arrangement: 0-4-0. Gauge: 3-feet 6 inches.

The first purpose built New Zealand locomotive. Placed in service in December 1872.
Eight months after the line closed, the *Evening Star* on 5 July 1875 reported serious storm damage to wharves and the coast line. The report mentioned damage to the Grahamstown Wharf and went on to say:

> A serious breach was made in the sea wall at the head of the wharf, the immediate result of which was to cause the engine shed of the Grahamstown and Tararu Tramway Company to tumble in two. The line of rails was so undermined that the two locomotives were in danger of falling over. They were, however, secured by ropes, but the place is now a great wreck.

The *Thames Advertiser* on 5 August 1876 reported a company in Napier was being formed to purchase the tramway and rolling stock of the Grahamstown and Tararu tramway, to lay tracks down Carlyle Street and Hydrabad Road to the Spit, about 2 miles: "the road is dead level, and would effect a great reduction in the cost of cartage." The newspaper made further comment:

> As the plant is lying in idleness at present, the owner will no doubt be glad to get rid of it. We believe it was intended to make it useful between the Puke Landing and Paeroa next summer if there was sufficient inducement.

The New Zealand Herald of 13 October 1880 reported:

> Two of the passenger carriages formerly used on the Tararu railway have been brought up from the Thames. They have been taken to Whangarei, where they will be used on the Kamo-Whangarei line. One of them was shipped by the S.S. *Waitaki*, and the other by the S.S. *Argyle* last evening.

Two-passenger cars and 12 wagons were sold to the Paeroa horse tramway in the 1890s – see page 42. *Some historians maintain the Fraser and Tinné locomotive was sold to the Kare Kare Timber Milling Company on the west coast near Auckland where it worked until about 1886. The remains of a locomotive were later found buried under sand at Kare Kare. This is not neccessarily the Tararu locomotive.
**The *Auckland Weekly News* of 19 February 1881 featured a story on the New Zealand Timber Company's property and operations on the Coromandel Peninsula which covered about 25,000 acres:

> The locomotive which is to be used on the railway is the largest and newest of the two formerly in use on the Tararu and Grahamstown tramway at Thames, and it is now undergoing a thorough overhaul and refitting by Mr T.T.Masefield. It is to be shipped next week, and will at once be set to work to enable the contractors to complete the plate-laying and ballasting of the line.

*See *New Zealand Steam Tram Engines and their Origins* by Ron Grant, page 191.
** From the Alan Bellamy archives

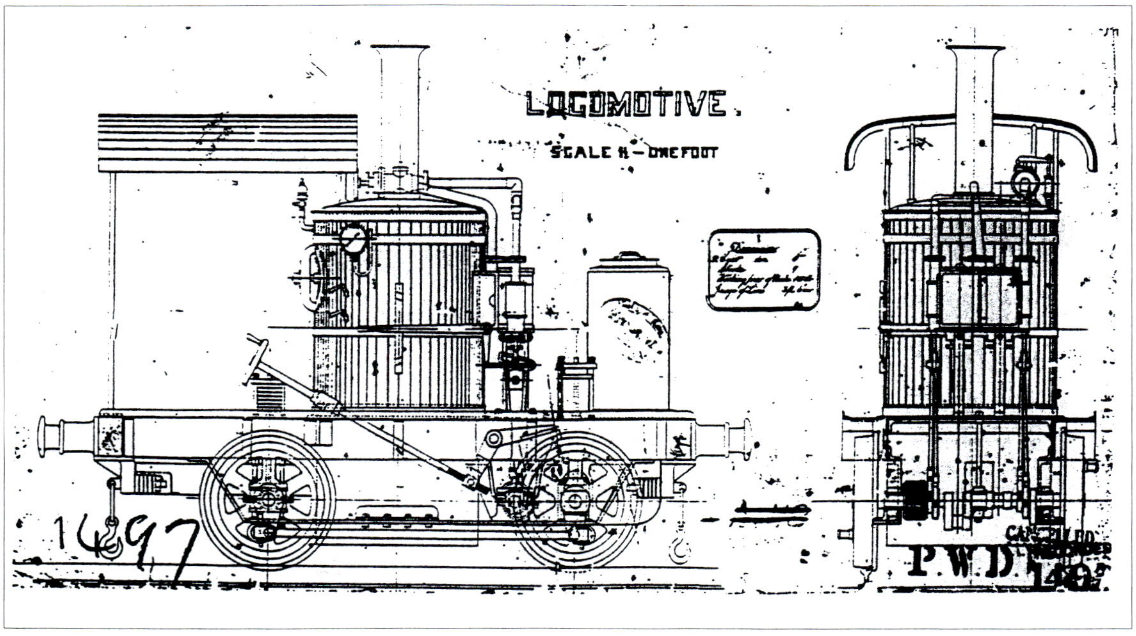

This drawing of a Fraser and Tinné built vertical-boiler locomotive from the Public Works Department's archives is thought to be similar to the one built for the Grahamstown and Tararu Tramway Company in 1872. *Alan Bellamy collection*

OHINEMURI

Tramway Company Ltd.

(To be Registered)

NOTICE is hereby given that the Tariff from the

Junction Wharf to Paeroa Terminus,

OR TO CUSTOMERS DOORS AS REQUIRED, and Vice Versa, will be as follows until further notice :—

	s.	d.
General Merchandise, per ton..	2	0
(Machinery or Goods per parcel, weighing over one ton by weight or measurement, by arrangement.)		
Timber per 100 sup.		6
Parcels under 5lb weight ..		6
" over and up to 56lb ..	1	0
Passengers, Single	0	6
" Return	0	9

By Order
C. W. GREENWOOD,
Traffic Manager.

Above: An advertisement from the *Ohinemuri Gazette* which first appeared on 30 September 1896. Some of the "tariffs" had already been reduced from the first published charges of 29 August 1896. *Les Turnbull*

Mackay & Pratt,

AUCTIONEERS.

IMPORTANT NOTICE.

TO BUSH, MINING, & COUNTY CONTRACTORS.

The Whole of the

Tramway Plant

at present lying at the

RAILWAY STATION,

PAEROA.

will be

SOLD BY PUBLIC AUCTION

On

THURSDAY, DECEMBER 16

(At 11 a.m. at Railway Station)

WITHOUT THE SLIGHTEST RESERVE.

The Plant comprises about eighteen sets of Wheels, Axles, and bearings, one tram car and sundries

The Plant will first be put up in one lot, and if not sold will be sold in lots to suit purchasers.

EDWIN EDWARDS
(For Tramway Company).

PAEROA

Horse Tramway

Ohinemuri Tramway Company

From 1 September 1896 to March 1897?

Two horse-trams and 12 goods wagons

(purchased from the Grahamstown & Tararu Tramway, *Ohinemutu Gazette*, 29 July 1896)

Gauge 3 feet 6 inches

The population of Paeroa was not recorded when the tramway opened.

From the Junction Wharf, along Junction Road to Railway Street, to Puke Road, then Belmont and Normanby Roads to the Criterion Hotel – about 1½ miles.

*On the day the tramway opened, two passenger horse-drawn tramcars met the steamer *Waimarie* at the junction and carried 36 passengers into the settlement, an extra tram being used to carry the luggage. Ten days later crowds watched when the trams left the Royal Mail Hotel to take passengers to the SS *Paeroa.* It was reported the cars were lighted by lamps both inside and out. It was intended to provide cushioned seats for the comfort of passengers.

Left: The auction notice for the sale of all the tramway plant which appeared in the *Ohinemuri Gazette* on 11 December 1897. Only one tramcar was advertised for sale; what had happened to the other car? *Trevor Terry collection*

Above: This photograph taken in 1896/7 looking north-west down Belmont Road, Paeroa, shows clearly the horse-tramway track on the left side of the road. The hotel on the left is the original Royal Mail Hotel. Two and four-horse teams feature in the photograph and parked on the right without any horsepower is a stage coach. The photograph is thought to have been taken from the old fire bell tower. *Alan Bellamy collection*

The *Ohinemuri Regional History Journal*, issue number 37, September 1993, in an article written by well-known railway historian Trevor Terry, of Hamilton, gives an interesting researched account of this short-lived New Zealand horse-tramway.

* Trevor Terry research

WANGANUI Gonville & Castlecliff Tramway Board

Steam-Tram

Baldwin Type locomotive
Fleet number: 100 (the original number as designated by the NSWGT)
Year Built: Completed March 1891 for the New South Wales Government Tramways, Sydney, Australia.
Australian Service: Entered service September 1891 for the NSWGT. Sold February 1905 to Saywell's Tramway and Estates Limited – used as a stand-by locomotive on the Rockdale electric line.
Sold to Wanganui late 1910.
Service at Wanganui: 1911-1912 construction work;
30 July to 11 October 1920, power failure crisis – passenger service;
15 and 23 September 1950 for parade and special gala runs.
Withdrawn: In storage 1912 to July 1920, December 1920 to September 1950. Sold for scrap 1950.
Builder: Baldwin Locomotive Works, Philadelphia, USA.
Maker's number: 11665. Class 4-16-C-87
Specifications as built: Wheel arrangement: 0-4-0. Wheel base: 4 feet. Wheel diameter: 2 feet 11 inches. Cylinders: 11 inches x 16 inches. Stephenson valve gear. Boiler diameter: 3 feet 4 inches. Firebox: 3 feet 6 inches x 2 feet 10 inches. Tubes: 102 – 1¾ inches in diameter, 5 feet 4-1/8 in length. Steam pressure: 140 lbs. Grate area: 10 square feet. Heating surface: 287 square feet. Saddle tank capacity: 300 US gallons. Weight in working trim: 15 tons. Tractive force: 6100 lbs.

Purchased from Saywell's Tramway and Estates Limited, late 1910, by the contractors for use on the construction of the Gonville-Castlecliff Tramway Board's electric tram line to Gonville and Castlecliff. Used to haul wagons carrying sleepers and girder tram rails as the line was built from the port to Gonville. Also hauled ballast from the Polson Street pit near the Wanganui Abattoirs at Castlecliff.

Became the property of the Gonville and Castlecliff Tramway Board when construction was completed in 1912 and stored behind the Gonville Town Hall. Used one summer to heat the Gonville baths.
When the electric power supply failed on 20 July 1920, the Baldwin tram locomotive was placed back on the rails and used to haul electric tram trailers between the city, Gonville and Castlecliff from 30 July. A restricted electrc tram service started to Aramoho and Wanganui East only on Thursday, 9 September using four trams. When the electric trams started running a full service again on Monday, 11 October, the Baldwin was placed in storage at the back of the tram barn in Taupo Quay.
Thirty years later when the electric tramway service closed in September 1950, the Junior Chamber of Commerce arranged a final procession of tramway vehicles through Victoria Avenue, the main street of Wanganui. "Puffing Billy" was moved back on to the tracks and given a certificate to run on half the original boiler working pressure.
Then purchased by the Blake Engineering Company for scrap and moved to their premises in Trafalgar Place. In 1955 purchased by the Steam Tram and Railway Preservation Society of New South Wales, Australia. As the society could not finance the cost of shipping the tram locomotive back to Australia, it was purchased by the author and Mr Peter Mellor in 1958, and donated to the Old Time Transport Preservation Society at Matakohe, Northland. Moved on 4 March 1958 to temporary storage on a farm in Mosston Road, Wanganui. In July 1962 transported to Auckland and stored at the tramway workshops in Manukau Road, Epsom, then to MOTAT in 1971. Now fully restored and reboilered and used on special occasions as a live exhibit.

Left: The Baldwin tram locomotive built in 1891 for the New South Wales Government Tramways, working six ballast waggons on the construction of the tramway to Gonville and Castlecliff. *M.R.Creagh*

Right: Ex NSWGT No. 100 working a ballast train on the construction of the electric tram line and road to Gonville and Castlecliff in 1911. Road metal was being unloaded by labourers. *M.R. Creagh*

Wanganui Herald
29 July 1920:

Drawing three cars "Puffing Billy" was the subject this morning of exhaustive tests, made under the supervision of the inspector of machinery. He was taken over the bridge at Guyton Street, his brakes were tested, and at the conclusion of the trail run permission was given for the engine to take up its duties as a tramway tractor. It will maintain a service from the Post Office to Castlecliff, running a regular timetable. The motor buses now on that run will be diverted to other routes.

Right: Leaving the city terminal in Ridgeway Street with two electric tram trailers during the electric power failure crisis in August 1920.

Left: Crossing the overhead Guyton Street bridge during the 1920 emergency service to Castlecliff. *Tesla Studios*

Left: At Castlecliff terminus before an afternoon run to the city during the power breakdown in 1920. From left: G.R. Holmes, later tramways superintendent and manager of Greyhound Buses Limited, who stoked the locomotive; the driver, A. Oliver; and conductors W. Coleman and A. Randell. *Tesla Studios*

Left: In retirement – sitting on a short length of track by the Gonville Municipal Baths in 1918. F.J.Denton collection, *Alexander Turnbull Library, G21177 1/1*

Above: In storage at the back of the tram barn in Taupo Quay between 1920 and 1950. *W.W. Stewart*

Below: After 30 years in storage, No. 100 made a triumphant return to the streets of Wanganui in a Cavalcade of Transport parade to mark the change-over from trams to buses – 17 September 1950. *Tesla Studios*

Above & Right: No. 100 was used again on Saturday, 16 September 1950 on excursion runs in Victoria Avenue, from the bottom of the avenue to the loop just past Guyton Street. The locomotive reversed the two trailers down the avenue after each trip to Guyton Street during the three hours the service ran. *Graham Stewart*

Left: In the hands of a scrap merchant – outside Blake Engineering in Trafalgar Place, Wanganui, before being rescued for preservation. *Graham Stewart*

Above: After complete restoration by the tramway division of the Museum of Transport and Technology, Auckland, in April 1996. *Bruce Pullar*

Left: In temporary storage on a farm in Mosston Road, Wanganui, 1958. Mervyn Sterling (left) and the author inspecting the relic. *Ian Stewart*

WELLINGTON

Steam & Horse Tramway

The Wellington Tramway Company

24 August 1878 to 1904

Company sold by auction to Mr E.W. Mills on 24 March 1880.
(Horse-tram operation replaced the steam trams in 1882.)
1880s – Sold to Dr M.S.Grace, Messrs J. Greenfield and C. Stewart.
Leased to Mr Archibald Hall from 1890 to 1898.
then to Dr M.S. Grace from 1898 until purchased by
the Wellington City Council in 1900.

Wellington City Council

1900 to 1904

Gauge 3 feet 6 inches

The population of Wellington was 19,000 when the system opened.

The Suburbs and Streets Served by the Steam and Horse Trams

Along Lambton Quay, Willis Street, Manners Street, Cuba Street, Vivian Street, Cambridge Terrace, around the Basin Reserve, to terminus in Adelaide Road, at the corner of King Street 24 August 1878.
Branch line from Lambton Quay, along Grey Street to the entrance of Queen's Wharf August 1878.
Thorndon Quay to Pipitea Point (just south of Davis Street) 5 September 1878.
Extension to Newtown 10 November 1879.
Via Courtenay Place line March 1881.
Loop line along Featherston Street September 1881.

Evening Post 15 March 1887:

> The proprietors of the City Tramways have just imported from the manufacturers, the John Stevenson [*sic*] Company, of New York, two cars constructed on an improved principle. In general appearance the new cars resemble those already in use, but several improvements in matters of detail have been introduced which will add materially to the comfort of the travelling public. They are fitted with spiral springs, in addition to those of gutta percha, the effect of which will be that in passing round curves a gentle rocking motion will take the place of the unpleasant jarring that is sometimes experienced. The windows are encased in gutta percha, in order to prevent rattling; the seats are provided with more comfortable backs than those of the other cars; the bell-pulls are of leather, instead of wire; the brake arrangements are less complicated than usual; and in many little ways the new cars have advantages which will be appreciated by those who have occasion to patronise them. One of them, if not both, will be placed on the line for the first time on St. Patrick's Day.

Wellington Tramway Company

Steam-Trams

Merryweather Type (Eight tram locomotives)
Fleet Names: *Florence, Wellington, Hibernia, Zealandia, Victoria, Anglia, Scotia, Cambria.*
Year Built: 1878. Last tram locomotive to be delivered placed in service 8 November 1879.
Withdrawn: January 1882
Builder: Merryweather & Sons Limited, London, England. Cost £975 each.

Fleet Number	*Name*	*Builder's Number*	*Arrived Wellington*
1	*Florence*	60/1877	1/7/78
2	*Hibernia*	61/1877	5/7/78
3	*Wellington*	62/1877	13/7/78
4	*Zealandia*	63/1877	22/8/78
5	*Victoria*	64/1877	/10/78
6	*Cambria*	85/1878	/11/78
7	*Scotia*	87/1878	21/12/78
8	*Anglia*	86/1878	31/12/78

Specifications as built: Wheel arrangement: 0-4-0. Wheel base: 4 feet 6 inches. Length over buffers: 6 feet 7 inches. Wheel diameter: 2 feet. Cylinders: 7 inches with 11-inch stroke. Firebox: 2 feet 2 inches x 2 feet area. 4.33 square-feet of grate. Firebox surface: 24.5 square feet. 79 flue-tubes, 1¾ diameter outside, 3 feet 6 inches long. 126.6 square feet of heating surface. Total heating surface 151.1 square feet. Water tank condensers on the roof. Diameter of barrel of boiler 2 feet 6 inches.

Above: No. 1 steam-tram locomotive *Florence* of the Wellington Tramway Company on opening day, 24 August 1878. *James Bragge, Museum of New Zealand; William Main, detail*

Three tram locomotives, the *Zealandia, Hibernia* and *Wellington*, were used to officially open the service on Saturday, 24 August 1878.

Steam operation ceased in January 1882 and the remaining tram locomotives sold. The *Anglia* and *Scotia* had already been sold to the Dunedin, Peninsula & Ocean Beach Railway Company Limited in December 1880.

Of the remaining six locomotives, one was retained to drive a chaffcutter for the horse-trams (either the *Zealandia* or *Florence*), the *Hibernia* was purchased by the Foxton-Sanson Tramway in 1884 who on sold it to a flaxmiller, E.S.Thynne who used the engine for driving the mill's machinery on the banks of the Rangitikei River at Parawenui, near Bulls. According to Ron Grant, the *Hibernia* was lost when the river flooded and is probably still buried under the river shingle. The *Wellington* went to a sawmill in Taranaki, the *Victoria* to the Tamaki Sawmilling Company in Woodville in 1886, then on sold to the Napier Harbour Board in 1896, becoming NHB No. 2, and the *Cambria* to the New Zealand Timber Company, which became the Kauri Timber Company in 1883. The *Zealandia* was sold to the Kauri Timber Company to haul logs (date unknown) and the *Florence* to the Kauri Timber Company in 1897.

Above: Merryweather-built steam-tram *Victoria* which was given the fleet No. 5 by the Wellington Tramway Company.

Above: *Florence* and *Hibernia* with trailers built by John Stephenson and Company, New York, USA, on opening day at the Adelaide Road depot on the corner of King Street. *James Bragge*

Left: *Zealandia* after the water tank condensers had been removed from the roof. *P.C. Sorrell*

Evening Post
6 September 1878:
The engines and cars ran along the tramway the whole distance to the railway station yesterday afternoon for the first time. On this occasion also two cars were taken, one long and one short... The fourth engine, *Zealandia*, commenced running this morning...

Left: No. 3 steam-tram *Wellington* with trailer No. 3 in Lambton Quay outside the Government Buildings in the 1880s. *Alexander Turnbull Library F8129½*

Steam-Tram Trailers / Horse-Trams

Fleet Numbers: 1 to ? Known numbers, 1 to 8
Year Built: 1878
Withdrawn:
Builder: John Stephenson Company Limited, New York, USA.
Specifications as built: Single-truck, single saloon with longitudinal seats, curved lower rocker panel, straight upper waist panel, open fronts. Stephenson uncushioned running gear. Bombay roof.

The *Evening Post* on 11June 1878 reported the completion of the lines and that a Government engineer would inspect the tramway within a day or two. 18 tramway cars built by John Stephenson for the Wellington Tramway Company had been shipped from New York via Dunedin on the bark *G.M.Tucker* and were "considerably overdue", as were the arrival of the first engines. The first engine, the *Florence,* arrived on the *Pleione* from the UK and was unloaded on 9 July. The bark *G.M.Tucker* did not arrive until 31 July, as the cars had been loaded on top of all the other cargo and had to be taken out and then loaded again at Port Chalmers.

Above right: John Stephenson built cars in service as steam-tram trailers in Lambton Quay.

Right: No. 5 leaving the Adelaide Road depot on the opening day of the steam-tram service in August 1878. *James Bragge, Museum of New Zealand, William Main, detail*

Evening Post
8 October 1878
Theatre-goers will be pleased to learn that extra tramway cars will be waiting at the Theatre Royal this evening, and through the week, to convey passengers to Te Aro and Adelaide Road after the performance is over.

Right: No. 8 being hauled by Merryweather steam-tram *Victoria* in Lambton Quay.

Above: Stephenson-built cars after the steam-trams had been withdrawn and replaced with horse-power. These two-horse trams were waiting in Lambton Quay for the evening rush hour outside the the Government Buildings. *Alexander Turnbull Library 21681½*

Right: A Stephenson horse-tram of this series working on the Newtown service in the 1890s was covered with advertising.

Evening Post
8 October 1878

The new tramway engine *Victoria* took her place in the regular running this morning and worked very satisfactory. The *Florence* has been laid up for a rest and an overhaul.

Right: Being hauled by a Heisler steam locomotive, a former Wellington Stephenson horse-tram being used as a carriage on the Taupo Totara Timber Company's railway.

Horse-Trams

Fleet Numbers: Unknown. Known numbers 15, 17.
Year Built: 1880s
Withdrawn: Unknown
Builder: Possibly J.G. Brill and Company, Philadelphia, USA.
Specifications as built: Single-truck, double-deckers, single lower saloon with longitudinal seats, curved lower rocker panel, straight upper waist panel, open upper deck with knife-board centre back-to-back longitudinal seating, left-hand quarter-turn 90° staircases at both ends, open fronts. Stephenson patented cushioned running gear.

Above: No. 15 stops in Lambton Quay. When these double-deckers were placed in service , they had the saloon rocker panels – the lower curved panel on the side of the tram – sign-written with the words "Wellington City Tramway Carriage".

Above: Wellington City Councillors celebrated the purchase of the Wellington Tramway Company on 1 October 1900 by making a special journey over the line on this double-decker horse-drawn tram with Dr M.S. Grace who had held the lease. *J. H. Brown*

Above: No. 17 with three horses out front working on the Courtenay Place run in the 1890s.

Above: The same series of double-decker trams passing the now Bowen Street intersection in Lambton Quay on a run to Courtenay Place.

Above: Looking up Cuba Street from the Manners Street intersection. A five-window saloon built by John Stephenson of New York when the tramways opened.
Right: No. 19, originally two Stephenson built five-window saloons, had been joined together to make one large car.

Above and left: Four-compartment Palace horse-trams in service on the Courtenay Place route. Three of these horse-trams were later purchased by the Kelburne and Karori Tramway Company and converted for use as trailers on the cable tramway. They would have given the longest service of any vehicle built originally as a horse-tram in New Zealand, finally being withdrawn from service on 27 July 1974.

Right: A busy horse-tram scene with five trams all heading for the railway station along Lambton Quay in the 1890s. *Littlebury series, Trevor Terry collection*

Right: The sweeping curve of Lambton Quay when a single horse-tram track was the only fixture on the roadway. *St George collection, Alexander Turnbull Library F21408½*

Below: Looking down on Lambton Quay in the 1890s with The Terrace and Bolton Street in the background. A lone horse-tram waits for a passenger to board.

Above: A small saloon horse-tram stops in Grey Street before proceeding across the intersection into Lambton Quay in 1877 bound for the Railway Station. *Museum of New Zealand, Te Papa Tongarewa B009760*

Left: A double-decker horse-tram waits for a Palace horse-tram at the Hunter, Featherston Streets intersection in Lambton Quay. The Palace horse-tram would have turned right at Hunter Street to run down Featherston Street.

Left: A view from Hunter Street of the same intersection. The horse-tram in Lambton Quay is heading for Courtenay Place. *Alexander Turnbull Library 3830½*

WELLINGTON CITY TRAMWAYS.
NOTICE.
PENNY FARES,
STARTING MONDAY, 13th FEBRUARY.

From Newtown to Junction of Cambridge-terrace and Vivian-street ... ... ONE PENNY
From Junction of Cambridge-terrace and Vivian-street to Evening Post Corner, Willis-street ... ... ONE PENNY
From Evening Post to Pipitea Point... ... ... ONE PENNY
Return Fares Similar.

Tramway Bus from Botanical Gardens *via* Princess Hotel, Molesworth-street, to Government Buildings and back, PENNY FARES EACH WAY.
First Bus leaves Botanical Gardens Tuesday next at 7.55 a.m., last 6.35 p.m., in connection with the Cuba-street Trams.
Through tickets can be procured on this Bus to Newtown by Tramway.
Packets of 12 penny tickets can be procured on application, and used either for Bus or Tram.
The packets of 2½d tickets can be used for all through fares.
SYDNEY SMITH,
Manager.

Evening Post 13 February 1888

Right: A loaded Palace horse-tram in Lambton Quay at the Willis Street, Customhouse Quay intersection. Stewart Dawson's is on the left.

Below: A horse-tram in the distance turns into Manners Street from Willis Street. The gradient of Boulcott Street is in the background

Right: Two Palace horse-trams, both being hauled by three-horse teams, in Manners Street.

Above: Looking up Vivian Street in the early 1900s with two double-decker horse-trams approaching. *Daroux*

Left: Commuters transferring from horse-trams (on the left) to horse-buses in Cambridge Terrace in the 1900s.

Above: The depot at the corner of Adelaide Road and King Street (later Drummond Street), Newtown. *Upper photograph Alexander Turnbull Library F51900½*

The building on the left in the lower photograph was built in 1887 and housed a ticket office, a waiting room for guards (conductors), and the manager's office.

In Affectionate Remembrance

of the

Wellington Horse

Cars,

which succumbed to

an Electric Shock,

Aug. 18th, 1904.

After 26 years of

faithful service.

GONE BUT NOT

FORGOTTEN.

WELLINGTON

Cable Tramway

Kelburne & Karori Tramway Company Limited

From 22 February 1902 to 11 December 1946

Wellington City Corporation Tramways

From 11 December 1946 to 1 April 1951

Wellington City Corporation Transport Department

1 April 1951 to 22 September 1978
New system – 20 October 1979 to 1 July 1991

Wellington Cable Car Limited

A Council Company that owns the assets

The Operators

Harbour City Cable Car Limited
A joint venture company owned by East by West Ferries Limited
& Stagecoach Wellington Limited
1 July 1991 to 30 June 1997

Serco Wellington
1 July 1997 –

From Cable Car Lane off Lambton Quay, up a private right-of-way through three tunnels each about 4 chains long, to Upland Road, Kelburn. Stopping places at Clifton Terrace, Talavera Terrace, Kelburn Park (Salamanca Road). The cars cross at Talavera Terrace. Original system (1902 to 1978) was double track, 3-feet 6-inch gauge.

The new system (from 1979) single track, metre gauge, with a passing loop at Talavera Terrace, the midway point. Still in operation.

Kelburne & Karori Tramway Company

Cable-Trams

Numbers: 1 and 2
Year Built: 1901 – into service 22 February 1902
Withdrawn: 22 September 1978
Builder: Mark Sinclair, Dunedin.
Specifications as built: Four-wheel, cable grip tram, enclosed saloons with sliding doors with two cross-bench seats facing each other at either end. In the centre on each side, knife-board centre back-to-back longitudinal seating with the gripman's compartment between, clerestory roof. Length: 19 feet. Width: 7 feet 9 inches over steps. Height to top of roof: 9 feet 3 inches. Height to top of pantograph: 10 feet 9 inches. Wheel base: 6 feet 8½ inches. Diameter of wheels: 2 feet. Seating capacity: 32.
Designed from patterns of the Mornington cable-cars and cars built by Mark Sinclair for the Kaikorai Tram Company. Steel reinforcement added to frame members in 1974.

Above: One of the cable-cars in original condition in 1903 before the outside longitudinal seats were "stepped" so that passengers could have a comfortable ride on the 1 in 5.06 gradient without sliding down the seat. *A. P. Godber*

Left: No. 2 cable-car shortly after the line opened in April 1902 with the company name painted on the ends of the car. "Kelburne–Karori Tram Coy.Ltd."
Above: After the side longitudinal seats had been "stepped" and the former horse-trams shortened and placed in service as trailers.

Above: No. 2 running with trailer in December 1970 after 68 years of service. *Graham Stewart*
Right: No. 1 being lifted on to a transporter the day after the original service finished on Saturday, 23 September 1978. *Graham Stewart*

Cable-Tram

Number: 3
Year Built: Unknown
Withdrawn: January 1975
Builder: Unknown.
Specifications as built: Four-wheel, cable grip tram, enclosed saloons with sliding doors with two cross-bench seats facing each other at either end and an open section at each end with one cross-bench seat facing outwards. In the centre on each side, knife-board centre back-to-back longitudinal seating with the gripman's compartment between, clerestory roof. Length: 21 feet. Width: 7 feet 9 inches. Height to top of roof: 8 feet 6 inches. Height to top of pantograph: 10 feet 9 inches. Wheel base: 6 feet 8½ inches. Diameter of wheels: 2 feet. Seating capacity: 30. This cable-tram had a smaller centre gripman's compartment compared with No's 1 and 2. Steel reinforcement added to frame members in 1974.

Evening Post 24 February 1902

The traffic on the Kelburne-Karori Tramway during its first three working days has been very heavy, about 5000 persons having been carried over the line since it was opened on Saturday morning. Over 2000 people went up the incline yesterday, and a considerable proportion of these made the further journey between the up-hill terminus and Karori, many of them patronising the line of buses which the company has put on to connect regularly with the tram cars.

Left: No. 3 working the incline in the early years with trailer attached.

Above: No. 3 with trailer at the Kelburn terminus in December 1970. The open front compartment panel which had originally been covered with only wire mesh had been panelled to give more weather protection for passengers. *Graham Stewart*

Right: No. 3 running solo after the then Ministry of Works had ordered the removal of the brakeless trailer cars, as they did not conform with modern safety standards, in July 1974. *Graham Stewart*

Cable-Car Trailers

Numbers: 4, 5 and 6
Year Built: Unknown
Three Palace-type horse-trams built for the Wellington Tramways Company in the 1880s. See page 53.
Withdrawn: 27 July 1974
Builder: Unknown
Specifications as built: Three Palace compartments with one cross-bench seat in each compartment with sliding doors on each side and an open platform at end each with a cross-bench seat. Rebuilt and shortened by the Kelburne & Karori Tramway Company – only two enclosed Palace compartments with open cross-bench seating at each end. Length: 14 feet 9 inches. Width: 7 feet 9 inches. Height: 8 feet 4 inches. Wheel base: 8 feet. Diameter of wheels: 2 feet. Seating capacity: 30.

Above and left: The former horse-trams working the incline as trailers in 1905. *Russell Grigg collection*

Below: The wire-mesh front aprons were later panelled. *W.W. Stewart*

Above: Trailer No's 4 and 6 (right) in the 1970s. Trailer No. 6 with no advertising, was being used then as part of the spare "set" with cable-car No. 3 – being only placed in service when either of the regular sets were out of service for maintenance. The front apron panels had been lowered on all the trailers to give passengers a better view. *Graham Stewart*

Cable-Trams

Numbers: 1 and 2 – second generation rolling stock
Year Built: 1979 – into service 20 October 1979
Builder: Habegger A.G., Thun, Switzerland.
Specifications as built: double-truck cable-tram, fully enclosed saloons with two air-operated sliding doors on each side of the car. The floor is stepped to retain horizontal platforms at the doorways and between the seats which are back to back. Interior, varnished hardwood and synthetic woodgrain panelling. Driver's compartment at each end, but car can only be controlled from the front end. There is an earthquake sensor in the control circuits to stop the cars in emergency. 30 seated, 70 standing.

The Swiss built cable-trams are now in their 18th year of service on the new metre gauge system which opened in 1979. *Graham Stewart*

The Kelburn cable tramway was built in 1902 by the Kelburne (as it was then spelt) and Karori Tramway Company Limited. The line opened on 22 February 1902 and within a few years three "Palace" horse-trams were bought from the Wellington City Council, rebuilt and shortened for use as trailers.

Although designed as a true cable tramway, the two cars were joined together by a second wire rope known as the tail-wire. This cable was independent of the main drive cable and ran round a drum at the top so that one car balanced the other. Only the descending car gripped the driving cable and, as it was pulled down the track, it hauled the ascending car by the tail-wire. Therefore the term "cable-car" was really a misnomer, as the tramway was technically a funicular worked on the counterbalance principle. But the trams retained and used the true cable-tram grippers and brakes.

In December 1946 the Wellington City Council exercised its right of purchase and the cable line became part of the services of the Wellington City Corporation Tramways.

After the Government licensing body, the then Ministry of Works, ordered the removal of the trailer cars for safety reasons in July 1974, which halved the seating capacity, the City Council fought hard to retain the original system to no avail. Finally, agreement was reached for the construction of a new cable-car system, a single track of metre gauge, with a crossing loop at the mid-point, Talavera station.

The original system closed on 22 September 1978 and the present system opened on 20 October 1979.

The same scene over 90 years apart of the Kelburn cable tramway from the Botanic Gardens. *Right:* Shortly after the trailer cars were placed in service in the early 1900s. *Trevor Terry collection*
Below: The modern cars of today, No. 2 taken in August 1997. *Graham Stewart*

NELSON

Horse Tramway

The Dun Mountain Copper Mining Company Limited

3 May 1862 to 15 May 1872

The known lessees were: Thomas Nesbitt – 3 May 1862 to 1863, James McConkey – 1863, William Weeden & Alfred H. Brind – 1866, Charles and Joseph Bray – 1866 to 1870, William F. Taylor – 1870, Stevenson & McGlashan – 1871, Thomas Newton – 1872.

Crewdson's City Bus

Moses Crewdson — June 1872 to June 1884

Jonathan Harle – June 1884 to 30 May 1901

Sold to Nelson City Council and dismantled in 1901

Three horse-trams – gauge 3 feet
The population of Nelson was 3,700 when the system opened.

The Suburbs and Streets Served by the Horse-Trams

From the corner of Hardy and Trafalgar Streets, along Hardy Street, Rutherford Street, Haven Road to the Port of Nelson.

Charles and Joseph Bray who leased the Dun Mountain bus from 1866 to 1870. *W.E. Brown collection, Nelson Provincial Museum*

Moses Crewdson who purchased the Port section of the Dun Mountain line and city bus in 1872. *Tyree collection, Nelson Provincial Museum*

Jonathan Harle, the last owner of the city bus from 1884 to 1901. *Tyree collection, Nelson Provincial Museum*

Dun Mountain Copper Mining Company Limited

Horse-Tram

Fleet Number: 1
Year Built: 1862
Withdrawn: 1881?
Builder: Keary and Company, Pitt Street, Sydney, Australia.
Specifications as built: Single-truck, underslung semi-elliptic springs, 18-inch diameter wheels, double-decker, single lower saloon, drop sash windows, longitudinal seats with cushions, straight lower rocker panel, straight upper waist panel, open upper deck with knife-board centre back-to-back longitudinal seating, open fronts. Access to upper deck by two end bulkhead steps at each end. Seating: 40.

Shipped on the steamer *Prince Alfred* to Nelson, arrived on 23 March 1862. Body built of cedar. The low ceiling and straight saloon sides (unusual in tramway design for the period) made the interior centre aisle of the saloon narrow and cramped for passengers.

Right: No. 1 City "Bus" of the Dun Mountain Copper Mining Company Limited, the first passenger service vehicle on rails in New Zealand, which began service on 3 May 1862. *Nelson Provincial Museum C2631, detail*

Right: No. 1 at the Port of Nelson terminus of the city bus in front of the Ship Hotel in the late 1860s. *Nelson Provincial Museum C1735, detail*

Crewdson's City Buses
Horse Tram

Fleet Number: 2
Year Built: 1870s
Withdrawn: 1901
Builder: Unknown.

Specifications as built: Single-truck, underslung semi-elliptic springs, double-decker, single lower saloon, drop sash windows, longitudinal seats with cushions, straight lower rocker panel, straight upper waist panel, open upper deck with knife-board centre back-to-back longitudinal seating, open fronts. Access to upper deck by two end bulkhead steps at each end.

Same design faults as No. 1, narrow centre aisle, although the saloon ceiling higher.

Left: No. 2 city bus which is thought to have been built for Crewdson's City Buses in the early 1870s. Similar in design to No. 1, although with more head room in the saloon. Could have been built by the same coach builder of Sydney who built the first tram, or the original design could have been copied by a local coach builder. *Nelson Provincial Museum C2631, detail*

Left: No. 2 in Haven Road outside the home of the proprietor Moses Crewdson who is pictured at right. The bus had been repainted with a different style of lettering and lining from the photograph above. The bus yard was behind the house on the right, the points leading off into the yard being under where No. 2 was standing. The driver was William Strange. *Nelson Provincial Museum C3426*

Crewdson's City Buses
Horse-Tram

Fleet Number: 3
Year Built: mid-1880s
Withdrawn: 1901
Builder: Gorrie and Sons, Bridge Street, Nelson.
Specifications as built: Single-truck, underslung semi-elliptic springs, double-decker, single lower saloon, drop sash windows, longitudinal wooden slatted seats, curved lower rocker panels, straight upper waist panel, open upper deck with knife-board centre back-to-back longitudinal seating, open fronts. Access to upper deck by two end bulkhead steps at each end. Cost: £250. English ash used for the frame, cedar panelling, kauri flooring. Footboard provided on each side of the car for top-deck passengers.

The *Colonist* of 12 April 1881 gave a detailed description of the elaborate colour scheme of No. 3:

> The ground colour is flake white, picked out in fine lines by ultramarine blue edged with gold, the large wide panels are painted carmine and ornamented in gold, whilst the small belt panel above is green lake, lined in white, and centred with gold ornaments, and we may add, that, with the exception of the large corners which have been transferred, all the ornamentation is by the hand of the coach painter. Below the large panels, on each side, the same artist has printed in original characters, the words 'City Bus' in lake green edged with gold, and shaded with dark and light grey, and the scrolls at either side of the letters are of gold cut up with burnt sienna. On the footboards to the 'knifeboard' seat are painted the name of Mr Crewdson. On the end guards are painted handsome scrolls in emerald green and black with gold border, and on other panels are painted spaniels and hounds.

Right: No. 3, the last tramway vehicle built for the Nelson to Port service in the mid-1880s by local coach builders Gorrie and Sons. The bow-shaped roof and the curved lower rocker panels on the saloon sides gave this vehicle the necessary aisle and head room which the two earlier trams had lacked. *Tyree collection, Nelson Provincial Museum 10293*

Left: No. 3 at the city terminus in Hardy Street about 1895. The Commercial Hotel is in the background. By now advertising panels were earning the proprietor extra revenue. *Tyree collection, Nelson Provincial Museum 27382/3*

Above: Mike Johnston of Nelson who has researched in depth this pioneer horse tramway, in his book *Nelson's First Railway and the City Bus*, feels this could be the last photograph of the city bus while still in service. This photograph was taken by Ernest Wilton outside his studio in Hardy Street in 1900 or early 1901. *Nelson Provincial Museum C1205*

Above: No's 2 and 3 were purchased by the Puponga Coal and Gold Mining Company for £5. 5s and £4. 5s, and shipped to Puponga in Golden Bay, where No. 3 (above) was used as an office at the mine, with the running equipment still in position. *Colin Loach collection*

Above: In July 1903 No. 3 was converted into a post office at Puponga. The running equipment was still attached to the body. *Nelson Provincial Museum* C3238

LEAVES NELSON A.M.	LEAVES NELSON P.M.	LEAVES PORT A.M.	LEAVES PORT P.M.
9.0	12.30	9.15	12.15
9.30	1.0	9.45	12.45
10.0	2.0	10.15	1.45
10.30	2.30	10.45	2.15
11.0	3.0	11.15	2.45
11.30	3.30	11.45	3.15
12.0	4.0		3.45
	4.30		4.15
	5.0		4.45
	6.0		5.15

This was the first passenger tramway line in New Zealand and ran between the city of Nelson and its port for 39 years – from 1862 to 1901.

It was originally the city section of a 13½-mile railway built by the Dun Mountain Copper Mining Company Limited to transport horse-drawn wagons carrying chrome ore from the mines near Dun Mountain to ships at the port of Nelson. The company leased the one-mile city section of the line for the conveyance of passengers to different operators, then following several years of poor mining results, the assets of the company were auctioned in 1872 and the city section of the line and the one horse-tram were sold to Moses Crewdson. Two further trams were introduced by Crewdson who sold to Jonathan Harle in 1884.

The Nelson City Council purchased the line and two horse-trams on 30 May 1901. Later that day the mayor and several councillors rode on the last fare-paying journey to the port and back. The line was then closed. The removal of the rails at a cost of £19.18s was completed by 6 July 1901.

Left: The illustrated timetable issued by Charles and Joesph Bray in late 1866. *Alexander Turnbull Library*

Below: The last city bus, No. 3, waiting in Hardy Street for departure time at the city terminus in the late 1890s. Across the road in Trafalgar Street is the Masonic Hotel.

This book covers only the passenger rolling stock used on this line. The book, *Nelson's First Railway and the City Bus* by Mike Johnston, published in 1996 by Nikau Press, Nelson, tells the full story of this first New Zealand horse tramway.

Above: No. 2 city bus heading along Hardy Street in 1885. The driver of the horse-tram is standing at right on the front platform where there was a small bulkhead seat for his use. The Hardy and Rutherford Street intersection is in the foreground. *Alexander Turnbull Library F3391½*

Left: In the foreground is the horse-tramway, steam-railway crossing by the intersection of Haven Road and St Vincent Street in the mid-1880s. This was a first for New Zealand, built in 1880 when the Government extended the Nelson railway to the port. In the centre is the City Bus yard which ran off the original Dun Mountain line in Haven Road, through a lean -to building beside the house and to the yard and stables bounded by Saltwater Creek. *Tyree collection, Nelson Provincial Museum 182230/3, detail*

Right: Looking along Haven Road and the Dun Mountain line to the port in the mid-1860s. *Nelson Provincial Museum C1641*

Below right: The same curve in Haven Road in the early 1880s. No. 3 city bus is nearing the junction of Russell Street with Haven Road. Major reclamation for the extension of the Government railway to the port had taken place. Haven Road had remained at the same width because the port tram line, which followed the original foreshore of Nelson Haven, would have needed to have been relaid to the right by the railway line. *Nelson Provincial Museum C3108*

Below: No. 1 at the port terminus outside the Ship Hotel in the early 1870s. The tram line was higher than the surface of Haven Road at this point. *Bett collection, Nelson Provincial Museum 224*

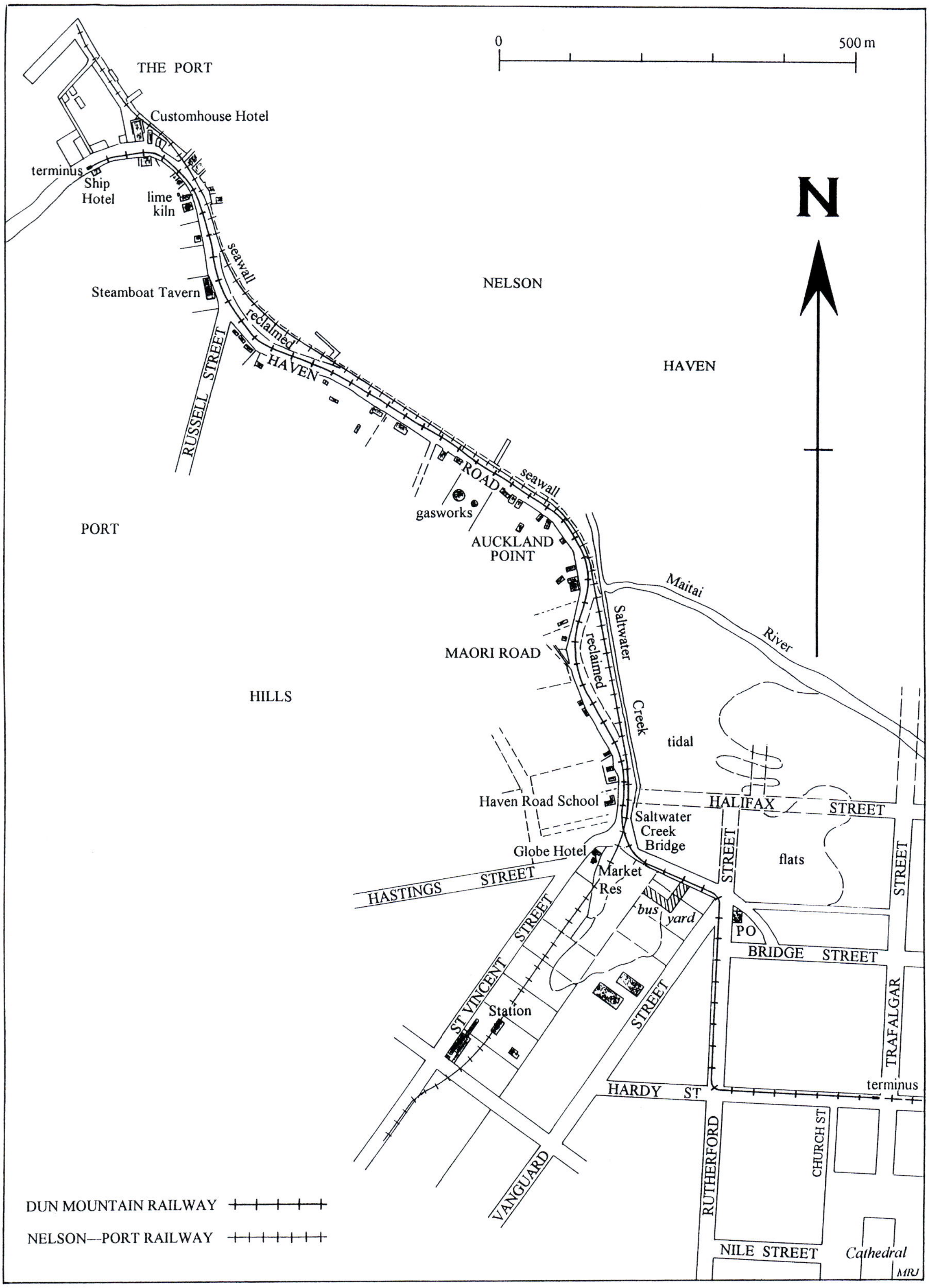

Left: The route of the city bus (Dun Mountain Railway) in the 1880s after the Government railway had been extended to the Port of Nelson. *Map drawn and by courtesy of Mike Johnston*

WESTLAND

Horse Tramways

All these passenger and freight lines were privately owned – wooden tracks.

Greymouth & Paroa Tramway January 1867 / The Saltwater Tramway Company 1868
The Greymouth & Kumara Tramway Company – 1877 to 31 December 1893

*Greymouth & Paroa (Saltwater Creek), 2 foot 6 inch gauge. Nearly 7 miles of narrow-gauge line, cheapest possible construction, mostly just above the high-water mark on the beach.
**One mile destroyed by the sea in October 1867 (company collapsed). A new company, the Saltwater Creek Tramway Company, rebuilt the line on a new alignment, extending it to Rutherglen in the south, and into Mawhera Quay, Greymouth, in the north. Much of the old line was rebuilt on a new alignment (solid ground), reopening in 1868. Length: 7 miles. This line eventually became part of the Greymouth and Kumara line.
This line extended to the Taramakau River in mid 1870s. In 1877 a steam-operated cableway (wire tram) was built to take passengers and freight across the river to a new tram terminus on the other side of the river by the site of the present Tramway Hotel – and a tram line, five miles in length, built to Kumara. Line closed when the Hokitika to Greymouth railway opened in December 1893.

Paroa to Marsden Tramway

**Another company built a line inland from Paroa to the Marsden diggings, near the present tourist Shantytown, in 1868. Length: 5½ miles.

The Hokitika & Greymouth Tramway Company

North Revell Street to Ho Ho Creek, December 1866, to Arahura River, 27 December 1866.

*Well constructed 4-foot gauge line from Hokitika to the coach road to Christchurch, terminating on the beach close to the mouth of the Arahura River. Built on the beach just above high-water mark. Cost £980 per mile exclusive of carriages and station buildings. Length: 4¼ miles. The Hokitika & Greymouth Tramway Company extended the line to Stafford in early 1867. Length: 8 miles.

The Hokitika & Kanieri Tramway Company

From Gibson's Quay, up the riverbank and through a cutting into the township of Kanieri.

Opened 25 December 1866.

*4-foot gauge line through forest – considerable earthworks and several bridges. Cost £1,038 per mile. Length: 130 chains. Altered and extended 2¾ miles.
**Continued until c1905 when it was taken over by a sawmilling company. About ¾ of a mile of this line still exists as a siding from the Hokitika Railway Station to the Westland Dairy Company factory.

Hokitika to Ho Ho Tramway

*4-foot gauge. Cost £946 per mile. Length: 2¾ miles. **Ran inland from Hampden Street, Hokitika, in an easterly direction to the Big Paddock diggings on the Ho Ho Creek. Opened June 1867.

The Ross & Hokitika Tramway Company

*5-foot 3-inch gauge. From the Ross diggings – beach track to Hokitika. Original line abandoned and a more direct line constructed in 1869. Tramway Street in Ross still exists.

The population on the West Coast was 28,400 in December 1867.

Westport had a horse-tram passenger service using trolleys with a back-to-back seat for pasengers running down Gladstone Street.
Charleston had a horse tram from the harbour, and Brighton once boasted horse-trams to St Kilda. Length: 1 mile. Opened 1867. St Kilda town has since been inundated by the sea. There was also the Donoghves Flat tram (near Ross), half a mile to the beach.

*From a report on these wooden-railed tramways from a paper 'Public Works in the Province of Canterbury, New Zealand' by E.Dobson, in the minutes of the Institute of Civil Engineers 1870.
** Ron Grant research.

Hokitika & Kanieri Tramway Company

Following the discovery of gold on the West Coast in the 1860s, horse tramways were constructed to link settlements along the coast and to inland towns that could only be reached through forest areas. With an abundance of timber, the laying of wooden rails was found to be cheaper than the cost of forming roads.

Above: A rural horse-tram at the Hokitika and Kanieri tram station with the proprietor, Tom McGuigan, seated in the saloon and his son on the driving platform. Kanieri was then spelt with two 'i's. *Alexander Turnbull Library*

Below: This photograph of the same or similar Hokitika and Kanieri Tramway Company horse-tram would have been taken in the winter months as all the canvas blinds had been lowered to give passengers protection against the weather. *West Coast Historical Museum*

Greymouth & Kumara Tramway Company

Right & Opposite page: Passengers travelling south from Greymouth and north from Hokitika and Kumara by horse-tram were carried across the Taramakau River by a suspended cableway cage. The wire tram, as it was known, was built in 1877 and was still in operation in the early 1900s. The photograph on the opposite page looks towards the north bank. *Trevor Terry collection*

Right: The Tramway Hotel Store and horse-tram station at the Taramakau River crossing. *William Heinz collection*

Below: Loading freight into the aerial cable cage which crossed the Taramakau River. On the right is the horse-tram from Kumara.

Above: The horse-tram terminus on the Kumara side of the Taramakau River. The wooden rails with track joints keyed and wedged without the use of nails are clearly seen in this photograph. *Burton Brothers collection, Museum of New Zealand Te Papa Tongarewa C001819*

Below: Another view of the horse-tram terminus at the Taramakau crossing. Two horse-trams can be seen at the end of the line. The horse-tram on the right carried freight only. The canvas cover is reminiscent of an American covered wagon. *William Heinz collection*

GREYMOUTH AND KUMARA TRAMWAY COMPANY (LIMITED).

TIME TABLE.

Leave Kumara, daily ... ... 8 a.m.
Do. ... ... noon.
Do. ... ... 4 p.m.
Leave Greymouth daily ... ... 8 a.m.
Do. ... ... noon.
Do. ... ... 4 p.m.

SUNDAY TRAINS.

Leave Kumara ... 8 a.m. and 4 p.m.
Leave Greymouth ... 8 a.m. and 4 p.m.

RETURN TICKETS—
TWELVE SHILLINGS,
Available for day only. Tickets issued on Saturday and Sunday available to Monday

W. HINDMARSH,
Secretary.
Greymouth, January 16th, 1678.

GREYMOUTH AND KUMARA TRAMWAY COMPANY.

NOTICE.

REDUCTION IN FREIGHT.

On and after Monday, the 24th September, AN ALLOWANCE of FIVE SHILLINGS per Ton of 2,000lbs. will be made on HEAVY GOODS and DEAD WEIGHT.

W. HINDMARSH,
Secretary.

Right: The Greymouth-Kumara horse-tram known to locals as the "bush tram". *James Ring, West Coast Historical Museum*

Below: The hotel-tram station at Paroa on the Greymouth-Kumara horse tramway. *West Coast Historical Museum*

Left: The Hokitika & Greymouth Tramway Company extension to Stafford and Goldsborough. The horse-tram ran from the coast at Awatuna to Pipers Flat for passengers and goods heading for the Goldsborough diggings south of Kumara. The sign across the side of the building in the left background reads "Pipers Tramway Station". *West Coast Historical Museum*

There were no roads and the tramways had a monopoly which even included tolls for pedestrians. Traffic was heavy and the returns to shareholders from these wooden tramway companies was in the order of 50 per cent per annum. The Ross line was the only venture that did not make money. The majority of the lines prospered for only a few years, being the life of each goldfield. *Many tramway companies had guaranteed agreements with the Westland County Council that no competing road would be constructed for X number of years, and some companies did claim compensation when a road was built in their territory and some were paid.

*Ron Grant research

Left: The last horse-tram about to leave Kumara for Greymouth on 31 December 1893. The stationmaster James Galbraith is standing in front. Even on the last run, passengers had to travel over the Taramakau River by the "Wire tram" to connect with the horse-tram to Greymouth. The service ended when the Greymouth to Hokitika railway was opened. *T. W. Young collection, Alexander Turnbull Library F 65470 ½*

CHRISTCHURCH

Steam & Horse Tramways

The Canterbury Tramway Company Limited

From 9 March 1880 to 14 April 1893 (In liquidation).
(Canterbury Tramway Company started horse-tram operation in April 1882.)

Christchurch City Corporation

23 April 1886 – Line leased to Mr Charles O'Malley.
August 1886 – Line leased to the Canterbury Tramway Company.
7 December 1888 – Line leased to New Brighton Tramway Company.
Sold in 1906 to the Christchurch Tramway Board.

The Christchurch Tramway Company Limited

Formed following the liquidation of the Canterbury Tramway Company on 26 June 1893. Sold to the Christchurch Tramway Board on 15 May 1905.

The New Brighton Tramway Company Limited

From 15 February 1887 – sold to the Christchurch Tramway Board on 1 August 1905.

The City & Suburban Tramway Company Limited

From 1 September 1893 to 1 October 1895 (In voluntary liquidation).
Purchased by Mr John Brightling and continued to trade under the same company name until sold to the Christchurch Tramway Board in November 1906.

Gauge 4 feet 8½ inches
The population of Christchurch in 1880 was 45,928 when the first line opened.

The Suburbs and Streets Served by the Steam and Horse-Trams

CANTERBURY TRAMWAY COMPANY / CHRISTCHURCH TRAMWAY COMPANY
Cathedral Square to Railway Station via Colombo Street 9 March 1880.
Cathedral Square to Railway Station via High and Manchester Streets 24 July 1880.
Cathedral Square to St Albans 5 June 1880, to Papanui 24 July 1880.
Cathedral Square to Sydenham Park 6 August 1880, to Devon Street 1 December 1880, to Cashmere (Dyers Pass Road) 7 December 1898.
From Manchester Street, along High Street, Ferry Road to Olliviers Road April 1882, to Woolston 7 July 1882, to Heathcote Bridge 9 December 1882, to Sumner via McCormacks Bay, Redcliffs and over a wooden bridge at Clifton Bay 1 November 1888, Nayland Street to Herberden Avenue January 1890.
Cathedral Square to Addington Railway Station via Colombo Street, Tuam Street, Hagley Avenue, Lincoln Road 5 January 1882, to Showgrounds October 1887, to Sunnyside (Hoon Hay Road) 1 January 1896.
CHRISTCHURCH CITY CORPORATION
From Oxford Terrace (where the Scott Statue now stands), Worcester Street, across the north side of Cathedral Square, along Worcester Street, Linwood Avenue (Canal Reserve), Buckleys Road, Slaughterhouse Road (now Rudds Road), to Night Soil Council Reserve March 1886. Branch line for hearse tram ran down Cemetery Road (now Butterfield Avenue), to Linwood Cemetery April 1886. Approximately 3 miles.

NEW BRIGHTON TRAMWAY COMPANY

From the corner of Rudds Road and Junction Street (now Coulter Street), along Pages Road as far as the Avon River 15 February 1887, Seaview Road to New Brighton Beach, George Street (now Oram Avenue), to stables August 1887. 2 miles 61 chains. The Canterbury Tramway Company operated the Christchurch City Council section of the line, from the Square to Rudds Road and the New Brighton Tramway Company their section of the line from Rudds Road to the beach from 15 February 1887 until the NBTC negotiated through running with the council on 31 December 1887. The New Brighton Tramway Company had constructed the line from Rudds Road to New Brighton Beach. The Square to Rudds Road line having been built for the Christchurch City Council.

CITY & SUBURBAN TRAMWAY COMPANY

From the intersection of Manchester, Lichfield, and High Streets, along Manchester Street, then Cashel Street, to Stanmore Road 1 September 1893, to Stanmore Bridge 2 September 1893, Stanmore Road, North Avon Road, North Parade, New Brighton Road to Burwood October 1893, through open country (now Bassett Street), to Travis Road, along Travis Road, through open country again (now Rookwood Avenue), to New Brighton Racecourse (now QEII Park) 20 March 1894, into Bowhill Road, to North Beach and along the Esplanade (now Marine Parade), to Seaview Road, New Brighton 25 October 1894. Length of the line 7½ miles.

Above: Cathedral Square in Christchurch Tramway Company days – a summer weekend with crowded double-decker trailers hauled by Kitson steam-tram locomotives. Passengers on the top decks of the trailers are standing to get a view of a street parade passing in front of the United Service Hotel at right. *Alexander Turnbull Library F84696½*

Much research was undertaken by David Hinman of Christchurch and Bruce Maffey of Hororata, on the Christchurch steam and horse-tram trailers for this book.
David Hinman is the Principal Policy Analyst for the Christchurch City Council; project manager, Christchurch tourist tramway 1993-95; former president, Tramway Historical Society; former Ferrymead Trust Board member; NZ executive member, Council of Tramway Museums of Australasia, and Canterbury branch chairman, NZ Historic Places Trust.
Bruce Maffey, is a genealogist; school teacher; former committee member of the Tramway Historical Society; a former editor of *Tramway Topics* and a well-known tramway historian.

Canterbury Tramway Company

Steam-Trams

Kitson Type (Eight tram locomotives) No. 1 standard type.
Fleet Numbers: Originally classified by the letters A to H. Then given the fleet numbers 1 to 8
Year Built: (Date of order) August 1879 – No's 1 to 5
(Date of order) May 1881 – No's 6 to 8
Withdrawn: No. 4 – 1893, No. 2 – 1925, No's 1 and 5 – 1928, No. 3 – 1944, No's 6 and 8 – 1952, No 7 – preserved.
Builder: Kitson and Company, Airedale Foundry, Leeds, England.
Maker's numbers: 1879 order: T8 to T12 (not necessarily in order)
1881 order: T27 to T29 (not necessarily in order).
No's 1 and 2 arrived at Lyttelton on 24/12/79 on the *Waitara*.
No. 3 arived at Lyttelton on 30/12/79 on the *Hereford*.
No. 4 arrived at Lyttelton on 16/01/80 on the *Piako*.
No. 5 arrived at Lyttelton on 09/04/81 on the *Gareloch*.
No's 6, 7, and 8 arrived at Lyttelton on 21/09/81 on the *Rangitiki*.
Specifications as built: Wooden cab. Wheel arrangement: 0-4-0T. Length over headstocks: 10 feet 4-5/8 inches. Width: 7 feet 6 inches. Wheel base: 4 feet 6 inches. Value gear: modified Walschaert. No's 1 to 5 – 6¼ x 12 inch stroke cylinders. No's 6 to 8 – 7¼ x 12 inches stroke cylinders. Wheel diameter: No's 1 to 6 –2 feet 3 inches, No's 6 to 8 – 2 feet 4¼ inches. Boiler barrel: 4 feet 3 inches long x 2 feet 5 inches in diameter. Water tank : 80 gallons. Firebox: 2 feet 6 inches long x 2 feet 10 inches in diameter. Tubes: 72 tubes 1½ inches in diameter. Heating surface: tubes 96.6 square feet; firebox, 19.4 square feet; total 116 square feet. Grate area: 5.17 square feet. Boiler pressure: 100 lbs per square inch. Weight in working order: 5¾ tons. Rating: 20 horsepower. Roof mounted surface condenser, coke fired.
The Christchurch Tramway Company reboilered the Kitson tram locomotives as follows:
No. 1 – 1894, No. 2 – 1898, No. 3 – 1893, No. 5 – 1893, No.6 – 1893, No. 7 – 1900, No. 8 – 1891. No. 4's boiler was condemned in 1893 and the locomotive scrapped for reasons unknown.
When reboilered, No's 1, 2, 3 and 5 received new cylinders 7¼ x 12 inch, and all seven locomotives had the wheel diameter increased to 2 feet 4¼ inches and the number of tubes reduced to 70. The working pressure was increased to 120 lbs. An additional 25-gallon tank was fitted. No 8's boiler had been condemned in 1888.
The seven remaining Kitson steam-tram locomotives taken over by the Christchurch Tramway Board in May 1905.
In the 1920s when the locomotives again required new boilers, three locomotives were scrapped, No 2 in 1925, No's 1 and 5 in 1928. No's 3, 6, 7 and 8 were reboilered in 1928. The cylinders were bored out to 8 inches x 12 inches, the diameter of the tubes increased from 1½ to 1¾ inches and the number reduced to 67. A live steam injector was fitted but the feed pumps were retained.

These steam locomotives were used for regular passenger services to Sumner, New Brighton via Burwood and North Beach, Riccarton, Opawa and Fendalton, by the Christchurch Tramway Board until replaced by the electric services, then used to haul ballast and pole wagons for new construction and track maintenance work until replaced by motor lorries in 1935, although still used on special occasions to haul passenger picnic trains. No's 3, 6, 7, and 8 were reconditioned in 1942 for emergency service should the power supply fail in the event of damage from enemy action and stored behind the Falsgrave Street car sheds. No. 3 was scrapped in 1944, and No's 6 to 8 towed to the Sumner shed and placed in storage. No's 6 and 8 were scrapped in 1951-52.
No. 7 was retained and used during the Canterbury Centennial Celebrations in 1950 for special gala runs to Sumner, New Brighton and the Riccarton Racecourse. Has since been retubed and restored by the Tramway Historical Society – used as a live exhibit at the Ferrymead Museum, Christchurch.

Below: Kitson No. 3 as built with trailer No. 12 in Canterbury Tramway Company days. *Tramway Historical Society*

Above: Two Kitson tram locomotives and Stephenson-built trailer No. 5 in the early days of the Canterbury Tramway Company. The second locomotive has the original classification displayed on the cab roof, the letter 'E' – later to be given the fleet number 5. The skirt around the wheel advertises Wertheim sewing machines on time payment terms from 2/6 a week.

Below: Canterbury Tramway Company Kitson No. 6 heading along High Street bound for Sumner, at the Tuam Street intersection. *Tramway Historical Society*

Kitson Steam-Tram Locomotives of Christchurch

Right: In the early years of the Christchurch Tramway Board, prior to electrification, No. 1 taking on water at the Burwood tanks while working the Richmond, North New Brighton Beach service. *Tramway Historical Society*

Right: No. 2 hauling four *Cage* trailers in Christchurch Tramway Board days.

Below: No. 3 shunting in the Square when owned by the Christchurch Tramway Company. *Auckland Institute and Museum*

Kitson Steam-Tram Locomotives of Christchurch

Left: No. 5 standing in the Square in the 1890s while double-decker trailers are being boarded by people bound for the beach.

Above: No. 6 waiting at the Sumner terminus ready to take commuters back to the city in the early 1900s. *Canterbury Museum*

Left: No. 8 photographed at the Sumner terminus with the driver's wife, a Mrs Taylor. Mr Merritt is the conductor on the left. Christchurch Tramway Company era. *Tramway Historical Society*

Christchurch Kitson Steam-Tram Locomotive No. 7

No. 7 has been restored by the Tramway Historical Society of Christchurch and is now the only surviving Kitson steam-tram locomotive in the world in working order. The three photographs above show the tram engine hauling closed saloon and open *Cage* double-decker trailers in Christchurch Tramway Company days.

Below: Still hard at work in the electric era, passing *Yank* No. 13.

Above: No. 7 back in service during the Canterbury Centennial Celebrations in October 1950 on the Sumner line, seen passing a timetable service electric tram and trailer. *W. W. Stewart* *Below:* As an exhibit at a Transport Exhibition during the London to Christchurch air race in October 1953 alongside NZR steam locomotive F13. *W. W. Stewart*

Canterbury Tramway Company
Steam-Tram Trailers / Horse-Trams

Fleet Numbers: 1 to 9 (nine cars)
Year Built: 1879
Withdrawn: Taken over by the Christchurch Tramway Company in 1893.
Builder: John Stephenson and Company, New York, USA.
Specifications as built: Single-truck, double-deckers, single lower saloon with longitudinal seats, curved lower rocker panel, straight upper waist panel, open upper deck with knife-board centre back-to-back longitudinal seating, left-hand quarter-turn 90° staircases at both ends, open fronts. Seating: 50. Length: 25'6". Length over body only: 16'0". Platforms: 4'9" x 5"0". Height: 10'3". Width: 5'9".
All Stephenson-built cars had bulkhead windows either side of the saloon doors.

The *Lyttelton Times* of 13 October 1879 published a detailed description including the following:

> There are 16 side lights and Venetian shutters, a metallic spring being attached to each side to keep the windows from rattling when the vehicle is in motion. The seats run from end to end in the interior in the same way as those in the railway cars at present in use on the Christchurch lines, but are constructed with a greater regard to comfort.
> The seats and backs are of American ash, perforated in tasteful patterns, the perforation being so arranged as to act as ventilators as well as a means of ornamentation. The walls of the carriage terminate in an 'interrupted arch' which allows for sitting room on the roof. Ventilation is provided by means of 36 small hinged windows under the outside seats. Running along the centre of the roof is a leathern strap, at each end of which is attached a spring bell for communicating with the guard. A polished handrail is placed on either side of the strap for the guidance of the 'unsteady'. The walls are of varnished ash and the roof is panelled alternately with white and bass coloured wood. In one corner of each end of the car is a neat lamp box lined inside with corrugated silver glass and fitted with a kerosene lamp. The two other corners contain polished mirrors which reflect the glare of the lamps opposite and intensify the light. Before leaving the interior it may be mentioned that the side seats are sufficiently far apart to give ample room to walk between two rows of passengers without inconvenience to anyone. The body is closed with sliding doors. The roof seats are approached by a spiral stair case of seven steps at each end of the car. The steps are covered with corrugated india rubber which, while it deadens the sound, renders slipping impossible. The roof seats run down the centre and are on the same principle as those of the ordinary omnibus. An iron railing some two feet high runs round the outer rim of the roof, which is covered with canvas and protected from injury by narrow boards or slats. The finish of the outside of the car is in keeping with the inside. The sides and ends are in broad panels of green and white and contain some small neat landscape paintings. The coupling rods and brakes are not yet attached to the carriage but we are informed that both are on an improved principle. The carriages have been built by Messrs. John Stephenson & Co. of New York and are very complete and comfortable in every respect. The packing is a marvel of ingenuity and the only damage done to the cars in their transit to Christchurch was the breaking of a few panes of glass, occasioned by the landing of the heavy packages in Lyttelton. The Directors of the Company may be fairly congratulated upon the effectiveness of their rolling stock.

Left: American-built Stephenson trailer No. 6 with No. 1 Kitson tram locomotive. *Alexander Turnbull Library F19743½, detail*

Below: Canterbury Tramway Company trailers No's 7 and 3 before modesty boards were fitted on the top deck. *Alexander Turnbull Library F38838½*

Canterbury Tramway Company
Steam-Tram Trailers / Horse-Trams

Fleet Numbers: 10 to 16 (seven cars)
Year Built: 10 – 1880; 11 to 16 – 1881/82
Withdrawn: Taken over by the Christchurch Tramway Company 1893.
Builder: W. Moor & Sons, Christchurch.
Specifications as built: Copies of the John Stephenson 1 to 9 cars. Single-truck, double-deckers, single lower saloon with longitudinal seats, curved lower rocker panel, straight upper waist panel, open upper deck with knife-board centre back-to-back longitudinal seating, left-hand quarter-turn 90° staircases at both ends, open fronts. Seating: 50. Length: 25'6". Length over body only: 16'0". Platforms: 4'9" x 5"0". Height: 10'3". Width: 5'9".
All Moor & Sons-built cars had bulkhead windows either side of the saloon doors.

*In November 1880 the company took delivery of a new trailer built locally to the John Stephenson design by Moor and Son and was given the fleet number 10. On 31 December 1880 the *Lyttelton Times* reported:

> It is satisfactory to note that the locally made carriage constructed for the Tramway Co. by Messrs Moor & Son has stood the very severe tests to which it has been subjected during the past two months extremely well, showing no sign of weakness, and appears in every respect equal to the imported cars. The carriage was weighed the other day, as it was supposed that owing to English ash having been used in the making in lieu of the American ash used in the imported carriages, it would be considerably heavier. It was found however, only to exceed the weight of the American cars by about 1cwt and ¾, a difference very much less than was anticipated.

* See *The End of the Penny Section*, 1993 edition, page 126

Above: Trailers No's 10 and 16 in the Christchurch Tramway Company era. No. 10 was the first street tramcar to be built in New Zealand, in 1880.

Right: Kitson No. 1 with one of the Moor and Sons-built trailers outside the Christchurch Anglican Cathedral. *Alexander Turnbull Library F91149½ Detail*

Below right: No. 11, built by Moor and Sons of Christchurch. The local builders announced with pride their coach-building skills on the letterboard of the car, written in capital letters "Tram-car and W. Moor & Sons Coach Builders ChCh" *Burton Brothers collection, Museum of New Zealand Te Papa Tongarewa C11575 Detail*

Canterbury Tramway Company

Steam-Tram Trailers / Horse-Trams

Fleet Numbers: 17 to 19 (three cars)
Year Built: 1882
Withdrawn: Taken over by the Christchurch Tramway Company in 1893.
Builder: W. Moor & Sons, Christchurch.
Specifications as built: Single-truck, single closed saloon with seven windows each side of the saloon, end bulkhead windows either side of the saloon doors, bow roof, open fronts. Designed to be pulled by one horse.

Above: No. 19 in the depot yard at the Canterbury Tramway Company's tramsheds in Cathedral Square. *F Bradley & Co, Canterbury Museum 2212, detail*

Three single-truck single deckers were built by W. Moor & Sons in 1882; refer the *Lyttelton Times* 20/11/82.

David Hinman from his research states that these three cars were converted into saloon double-deckers by the Canterbury Tramway Company and were listed as such in the 1893 schedule of assets when the Christcurch Tramway Company took over the Canterbury Tramway Company rolling stock. This accounts for some of the double-decker saloons with only seven windows which were later taken over by the Christchurch Tramway Board. (also refer New Brighton Tramway Company)

The company name was changed to W. Moor & Company in the 1884-5.

Above: Single-decker No. 18 standing alone in the Square, said to have been taken in 1885. A.*E. Preece. Canterbury Museum 8614, detail*

Above: No. 18 after being rebuilt into a double-decker. In Christchurch Tramway Company days, the bulkhead windows and two saloon windows on each side of the car had been replaced with solid panels.
John Bettle collection

Above: A Kitson steam tram hauling three trailers in Christchurch Tramway Company days, about to cross the Heathcote Bridge. The trailers are from left: No's 14 (Moor & Sons), 17 (Moor & Sons), and 2 (Stephenson). The seven window No. 17 in the middle, which started service as a single-decker, shows clearly the difference in design from the other trailers. *Bruce Maffey*

Canterbury Tramway Company
Steam-Tram Trailer / Horse-Tram

Fleet Number: 20 (one car)
Year Built: 1883
Withdrawn: Taken over by the Christchurch Tramway Company 1893.
Builder: W. Moor & Sons, Christchurch, built the first of this design as confirmed in *The Press* 30/07/83.
Specifications as built: Single-truck, open toast rack with cross-bench seats, entry/exit running boards, arch roof, glass end bulkheads, canvas weather blinds, open fronts.

The Press reported on 30 July 1883 the introduction of the first of this class:

> The latest addition to the rolling stock of the Canterbury Tramway Company is a somewhat ponderous looking open carriage built by Moor & Sons of Christchurch to accommodate and seat 40 passengers. It is made of iron with wooden flooring and is supplied with the most improved American machinery. In appearance it is decidedly ugly, presenting more the appearance of a large flat bottomed boat, than a passenger car. It is already christened the *Punt* and it certainly deserves the appellation. It has not yet been decided when the *Punt* will be brought into service, but as a smoking car for gentlemen, perhaps the sooner the better.

Right: This photograph taken in the Square in 1888 shows an open toast-rack tramcar in the foreground. The design is different from other open cars built in the 1890s and is thought to be Canterbury Tramway Company's No. 20, built locally by W. Moor & Sons in 1883. *Canterbury Museum 4874*

Right: This former Christchurch tramcar body was discovered in Flockton Street in 1971. *One theory is that it was the original Canterbury Tramway Company "Punt" No. 20, converted to a closed car. *David Hinman*

*David Hinman

Canterbury Tramway Company
Steam-Tram Trailer / Horse-Tram

Fleet Number: 21 (one car)
Year Built: 1885
Withdrawn: Taken over by the Christchurch Tramway Company in 1893.
Builder: Booth Macdonald, Christhcurch.
Specifications as built: Single-truck, open lower and upper decks with knife-board centre back-to-back longitudinal seating, decency boards, *left-hand straight staircase at one end only, open fronts. Appears to have had the other end staircase added later, perhaps by the Christchurch Tramway Company.
Known as a *Cage.*

* A new double-decker to carry 68. built by Booth Macdonalds, the extra amount of room resulting from access to the roof being provided by one staircase only. Dayle Maffey, *Tramway Topics*, August 1984, page 28

Above: This photograph taken in 1904 shows a special picnic tram chartered by the Waddington School at Sumner. The trailer behind Kitson steam locomotive No. 2 is No. 21 – the last vehicle built for the Canterbury Tramway Company in 1885 by Booth Macdonald of Christchurch. *Reverend T. Lyon, Canterbury Museum 14887*

Canterbury Tramway Company

Horse-Tram

Fleet Number: 51 (one car)
– re-built and re-numbered 43 in 1897/1898
– re-numbered 50 c1912
Year Built: 1887 (ordered August 1887) – arrived April 1888
Withdrawn: Taken over by the Christchurch Tramway Company in 1893.
Builder: John Stephenson and Company, New York, USA.
Specifications as built: Single-truck, single saloon, curved lower rocker panel, straight upper waist panel, clerestory roof, longitudinal seating, open platforms.

Rebuilt as a California combination by the Christchurch Tramway Company and given the fleet Number 43 in 1897/1898.
Re-numbered No. 50 by the Christchurch Tramway Board c1912. Never converted for air brakes. Centre aisle reduced seating capacity c1913.

The *Lyttelton Times* of 19 April 1888 described the one-horse tram:

> The Canterbury Tramway Company have almost completed the erection of a new imported one-horse tramcar which for general appearance and completeness is ahead of anything running in Christchurch. The new car fairly bristles with patents – patent india-rubber springs, to make it run smoothly, a patent lifeguard composed of two rods with rubber sheathing to prevent anything getting under the wheels. There is another patent to prevent the windows rattling, and yet another is a fare collector, an ingenious affair in glass, into which the passenger drops his contribution to the funds of the Company, after having been reminded to do so by the driver ringing a bell. This collection does away with the necessity for a conductor for the car, which will probably on its completion see service on the tram line to the Railway Station.

*The half-yearly meeting of the Canterbury Tramway Company held on 15 August 1888 reported the one-horse car from America was so highly ornamental, and had cost £100 more then expected, that the company would not use it until a dividend was paid.
It appears this tram was seldom used after the Christchurch Tramway Board took over the tramway services. Used in the "March of Progress" procession in December 1925, again during the "Youth Employment Campaign" in September 1932; it was then stored until 1950, when used at Sumner as a horse-tram during the Canterbury Centennial celebrations. Renovated by the Tramway Historical Society and used as a horse-tram in Papanui Road in 1964 during the 10th anniversary of the running of the last electric trams with the fleet number 43. Restored by the Tramway Historical Society in the 1980s to California style of the Christchurch Tramway Company era.

Above: The official photograph taken at the coach builder's factory in New York by John Stephenson and Company, New York. The single-saloon one-horse tram, No. 51, built for the Canterbury Tramway Company in 1887. The fleet number 51 remains a mystery – it should have been No. 22 in the Canterbury Company fleet. *Museum of the City of New York*

Above: A rare photograph of No. 43 in Cathedral Square, working as a horse-tram for the Christchurch Tramway Company. *John Bettle collection*

Above: No. 50 in the March of Progress procession in December 1925.

Above: Entering the Tramway Workshops in Moorhouse Avenue in July 1932 to be cleaned and repainted for the Youth Employment Campaign. *Green and Hahn, Bruce Maffey collection*

Above: Training the horse for the Youth Employment Campaign in Moorhouse Avenue in 1932.

Above: The birth of the Tramway Historical Society of Christchurch. The first restoration project undertaken by the society was the renovation of No. 50 which they renumbered 43, the fleet number given this car when owned by the Christchurch Tramway Company. The car gave free rides in September 1964 along Papanui Road during the 10th anniversary of the running of the last trams in the city. The tram tracks had not been removed or tarsealed over in 1964 and with a loop still in position, the car was ideal for the occasion. *Green & Hahn*

Above: No. 43 after being restored by the Tramway Historical Society in the 1980s to California style of the Christchurch Tramway Company era. *Colin Loach*

Canterbury Tramway Company
Water-Tank Trailer

Fleet Number: Unknown (one car)
Year Built: 1880
Withdrawn: Unknown
Builder: Canterbury Tramway Company
Specifications as built: Single-truck, two 400-gallon iron tanks.

No photographs have been found of this two 400 gallon iron tank trailer built by the Canterbury Tramway Company in 1880.

The *Lyttelton Times* of 15 October 1880 gave the following description of this water car with the heading "Watering the Tramway Line":

> Mr Peterkin the traffic manager of the tramways has devised an appliance for watering the line in order to remove the inconvenience caused during the dry weather by the clouds of dust arising through the passage of the tram. It consists of a strong wooden truck with iron trolley wheels bearing two 400 gallon iron tanks, which are connected underneath by a two-inch pipe. A pipe 6 feet long, perforated along its whole length is attached to the coupling pipe under the centre of the carriage by a brass connecting cock, by means of this pipe the whole width of the line and about a foot on each side of the rails will be thoroughly watered. A lever communicating with the cock, passes along the side of the carriage to the engine, so as to enable the driver to turn the water on and off. It is expected that the watering carriage will commence its work on Saturday next on the north line from the belt to Papanui. It will be drawn by a spare engine and will run up and down the line at intervals so as not to interfere with the traffic.

Below: This photograph of the city council yards, corner Worcester Street and Oxford Terrace, in 1900 shows one of the Christchurch City Council saloon double-decker trams and what appears to be the "dust trolleys" alongside. Refer opposite page. *Brittenden collection, Canterbury Historical Association colletion, Canterbury Museum, Detail*

Christchurch City Council
Horse-Trams / Two-Tram Trolleys / 12 Iron Carts

The council built this line for the purpose of carrying four types of loads: passengers, bodies, rubbish and nightsoil. Bodies and nightsoil were never carried.

In May 1885 tenders were accepted from Christchurch agricultural implement manufacturers Booth Macdonald for the construction of two double-decker passenger trailers and for a hearse tram by W. Moor & Co. These cars were delivered by March 1886.

June 1886 – Tenders called for two tram trolleys and 12 iron cart bodies for night soil cartage. The order went to P & D Duncan's foundry. Each cart was 5 feet 3 inches in length, 4 feet 2 inches wide. On top of the wooden frame were placed four tipcarts of 1/8 iron plate positioned so that the rubbish could be emptied alongside the line with ease.

The rubbish and dust service started in August 1886 and was worked between midnight and 5 am each morning. The rubbish tram service continued until the Municipal City Destructor was built in Manchester Street in 1902.

The length of the main line was 3 miles 8½ chains and including the sidings, the total length was 3 miles 20 chains. 40 lb rails were used.

Below: A busy day on the New Brighton Tramway Company's line. The four-horse team with two trailer cars had just arrived at New Brighton. The trailer car in the foreground has no saloon window areas covered with panels as was the design of the NBTC saloon double-decker cars. Could this car have been either Christchurch City Council No.1 or No. 2 which were leased to the New Brighton company? *John Bettle collection*

Horse-Trams

Fleet Numbers: 1 and 2. (two cars)
Year Built: 1885
Withdrawn: No.1 dismantled 1905, No. 2 taken over by the CTB.
Builder: Booth MacDonald, Christchurch.
Specifications as built: Single-truck, double-deckers, single lower saloon with longitudinal seats, curved lower rocker panel, straight upper waist panel, open upper deck with knife-board centre back-to-back longitudinal seating, left-hand quarter-turn 90° staircases at both ends, open fronts. Cost £245.00 each.

Leased to C. O'Malley, Canterbury Tramway Company, and from December 1888 to the New Brighton Tramway Company. When the Christchurch Tramway Board took over the assets of the New Brighton Tramway Company the CTB was not aware these two trailers were still being leased from the Christchurch City Council, as stated in a report to Council dated 4/3/07:

> I have to express regret that the fact that the Council is the owner of the two tramcars (1 & 2), formerly leased to the New Brighton Tramway company, was quite overlooked when delivery of that Company's plant was taken. The two cars mentioned were quite unfit for use when taken over, the condition being such that they were not even fit for repair. They were offered for sale at first for £10, ultimately reduced to £3, but were not sold. No. 1 was then dismantled, in the Board's yard, and No. 2 is still in the Board's yard at New Brighton. I am directed to say that the Board is willing to pay the Council what the cars would have been worth to the Council when the lease expired. If the Council is not satisfied with the value as indicated above, the matter might be settled by the Council selling the one at New Brighton and the Board paying the same price for the one that has been dismantled. Again regretting that the matter has been overlooked. *Signed Frank Thompson. Secretary, Christchurch Tramway Board.*

However, in a second letter to council dated 24 April 1907, the board wrote:

> Re Old Tram Cars. With reference to yours of March 1st last, I have to inform you that investigation shows that while No. 1 car was dismantled, No. 2 was not left at New Brighton as previously advised but was delivered to the Board by the New Brighton Tramway Company and was among those that were reconstructed by the Board. This car was in an exceedingly bad condition, and in fact the Traffic Superintendent was of the opinion that it was not worth repairing but at that time it had been included in a contract with Messrs Boon & Co for repairs to all trailers. In addition to the extensive repairs carried out by Boon & Co, new wheels had to be provided. I am directed by the Board to offer you the sum of £10 which it considers the fair value of the car when it was taken over. I regret that the Council's interest in these cars was overlooked.
> *Frank Thompson. Secretary, Christchurch Tramway Board.*

Christchurch City Council

Hearse Tram

Fleet Number: Unknown (one car)
Year Built: 1885
Withdrawn: July 1892
Builder: Moor & Co, Christchurch.
Specifications as built: Single-truck, single deck, platform at each end, elliptical plate glass windows, fine wooden panelling, painted black, railings on the low roof to hold floral tributes, could accommodate four coffins. Cost £225.00.

Built for the council who intended leasing the hearse tram to a private contractor to run a funeral tram service to Linwood Cemetery. The City Council thought it could provide low-cost funerals for less fortunate citizens. Never used used for a funeral service.

In January 1888 a proposal to convert the hearse tram into a standard trailer for passenger traffic was found to be too expensive; the cost would have been £120. **The Press* indicated that its last "public appearance" had been on the 28 July 1892 when it was drawn slowly through the Square and along the New Brighton line, during which journey passers-by showed a great amount of interest.

Dismantled and the body sold in 1901 for £3.00. Body used as a powder magazine on the mudflats near the Heathcote Bridge for Andrew's Quarry. Later placed on a pontoon and used as a houseboat on the estuary.

* Dayle Maffey, *Tramway Topics*, Febuary 1985

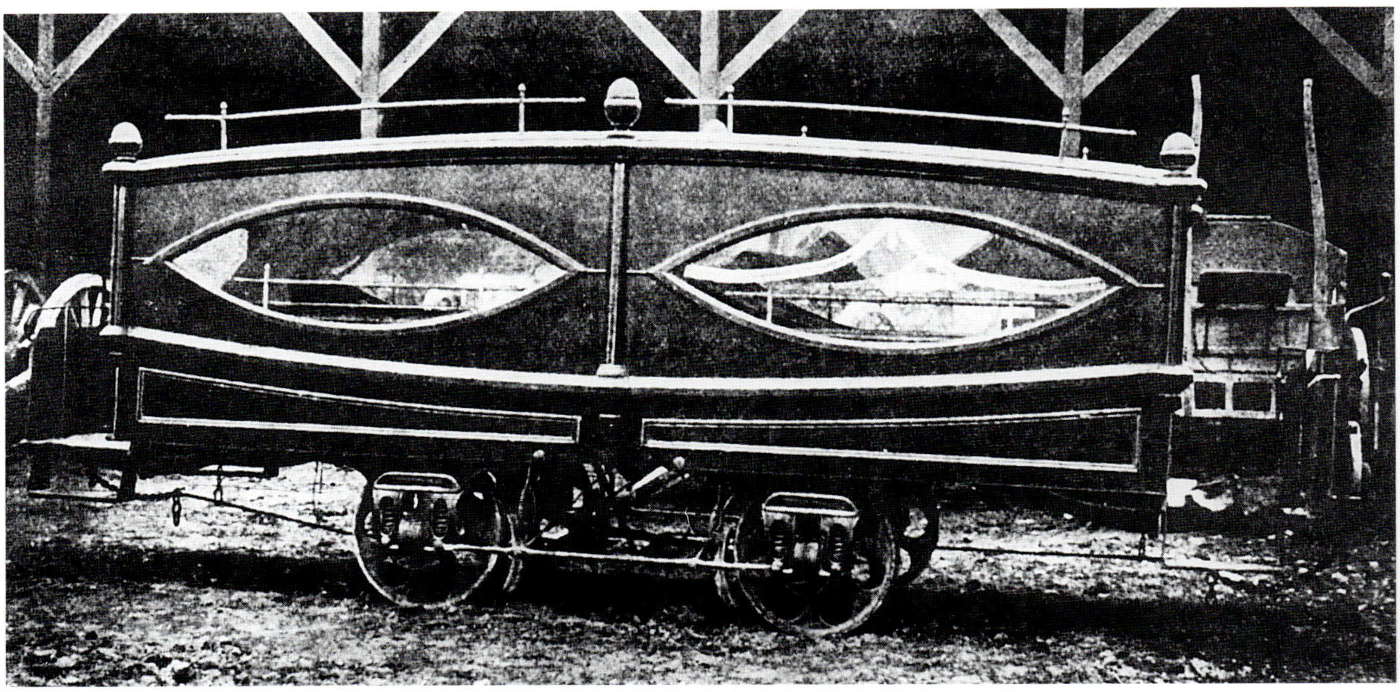

Above: Possibly the most unique tramcar to be built in New Zealand. The hearse tram which was built for the City Council to provide low-cost funerals for the less fortunate.

New Brighton Tramway Company

Horse-Trams

Fleet Numbers: 1 and 2 (two cars)
Year Built: 1885
Withdrawn: Purchased by CTB in 1905 as part of the company assets, but not added to the CTB fleet.
Builder: W.Moor & Co, Christchurch.
Specifications as built: Single-truck, open toast rack with cross-bench seats, entry/exit running boards, arched roof, glass end bulkheads, canvas weather blinds, open fronts.

*By 31 December 1886 one of the tramcars was complete and the other expected in a fortnight. A good description of these was given in the newspaper of that date:

> One of the two carriages to be used on the line is now finished by Moor & Son [*sic*]. It is somewhat of a novelty here, being what is known as an American Summer Car. The sides are open, but can be closed, if desired, by means of curtains of American duck. The ends are panelled and provided with windows. The roof, which is slightly arched, is supported on light wooden pillars, finished with ornamental brackets of cast iron. The seats, of wood, are placed across the car, which is entered from the sides. There are 10 seats, each of which will accommodate 5 persons comfortably. Total length of car is 23 feet and width 6 feet. Height from floor to roof is about 6 feet. The framework is constructed of American Ash, the floor and ceiling however being of native timber. The lower portion of the car is painted pea-green, and the seats and ends and roof pillars grained in imitation of oak and relieved by red lines and a small stencilled pattern of red and blue. The interior of the roof is of varnished kauri, while the exterior is painted white. The general appearance of the car is light and effective. Though the total weight is only about 35 cwt, the car is strongly and solidly constructed and should stand no small amount of wear and tear. The brakes and wheels are similar to those used on the cars of the Canterbury Tramway Company. It is worthy of note, however, that the whole of the ironwork, with exception of the wheels, is of local manufacture.

Known as *Punts* or *American Summer Cars.* There are no known photographs of NBTC No's 1 and 2.

Above: New Brighton Tramway Company No. 8 built by W. Moor and Co, heading for the city at a good pace with a four-horse team in control. See the following page for details.

New Brighton Tramway Company Horse-Trams

Fleet Numbers: 3 to 10 (eight cars)
Year Built: 3, 4 –1887; 5, 6 – 1889; 7 – 1892; 8 – 1893; 9, 10 – 1894.
Withdrawn: Taken over by the CTB in 1905.
Builders: 3, 4, 6, 8 – Moor & Co, Christchurch;
5, 7, 9, 10 – Boon & Stevens, Christchurch.
Specifications as built: Single-truck, double-deckers, single lower saloon with longitudinal seats, curved lower rocker panel, straight upper waist panel, open upper deck with knife-board centre back-to-back longitudinal seating, left-hand quarter-turn 90° staircases at each end, open fronts. No's. 5 and 6, seven window saloons, No. 5 built principally "for inside passengers" although "fitted to be used for carrying passengers on top", *The Press*, 16 April 1889.

*In May 1887 the company took delivery of a new double-decker tramcar constructed by Moor & Co (No. 3). A closed car seating 48 passengers and in general appearance resembled those used by the Canterbury Company except that the upper portion of the sides consisted of alternate wooden panels and glass frames, instead of being composed entirely of windows. The roof was supported by ribs of iron instead of wood as in the older cars. The ends were of angle iron giving considerable strength to the roof as well as durability and lightness to the entire structure. Ventilation was obtained by means of apertures in the roof boards instead of by small glazed frames as in the other cars. The apertures were protected by a "fence rail"outside, so as to prevent ingress of wind or rain. The roof and panels of the car were of kauri and the other framework of American and English ash. Inside seats were of perforated sugar maple. The painting was described as very effective. The power portion of the body was salmon coloured, striped with claret; the centre panels were painted in "Victoria Lake", a deep rich brown, and striped with vermilion and orange; the panels between the windows were cream colour, ornamented with a stencilled pattern in maroon. The dash was umber, striped with yellow and the staircase vermilion, also relieved with yellow. It was noteworthy that everything except the wheels had been constructed locally.
The *Lyttelton Times* of 22/10/87 reported on the arrival of what would have been No. 4 of the New Brighton Tramway Company's fleet:

> Messrs Moor & Company have just completed a new car for the New Brighton Tramway Company. A facsimile of that made for the same company about six months ago. Seats 50 passengers. Resembles the Canterbury Tramway Company cars, but differs in a few construction details. The framework is of English and American Ash, and panels and facings of colonial timber. The only imported parts are the wheels and the india rubber springs.

On 02/09/93 *The Press* reported on the workings of the New Brighton Tramway Company and the pending arrival of No. 8 being built by Moor & Co:

> ... A new car is well advanced towards completion. When it is finished the Company will be able to put on four-in-hand teams for holiday traffic. The cars have been throughly overhauled and painted and varnished, while additional purchase of horses have been made.

* Dayle Maffey, *Tramway Topics*, October 1984, pages 32, 33

Below: Three New Brighton Tramway Company trailer cars standing at the New Brighton terminus. Wheel guards had been fitted to the two cars nearest the camera. *Bruce Maffey collection*

A year later the *Lyttelton Times* reported the arrival of No's 9 and 10:

> New Brighton Trams – On Saturday the New Brighton Tramway Company received an addition to its rolling stock of two cars, which in many respects are superior to any yet turned out here. The work of building them was entrusted to Messrs Boon & Stevens, whose workmanship will be appreciated by the travelling public. A special trip was run by the company on Saturday for the purpose of trying the cars, and it was found that they ran perfectly smoothly, there being no vibration and an almost entire absence of noise. Upon arrival at New Brighton a luncheon was given by the contractors in Hawker's Hall, and the toasts of "The Tramway Company" and "The Builders" were honoured. The cars are hung on double spiral springs, with roomy platforms, a gutter round each end preventing any drips from the roof in rainy weather falling on the driver or on passengers using the platform. The doors are hung on a new principle, by which they are always wide open or shut close. They are also fitted with india rubber jambs and cushions, so that there is no objectionable rattle. The windows, which are set in cloth, are also provided with anti-rattlers. The finish inside is extremely good. At each end there is a bevelled glass mirror, while the corresponding panels are of huon pine with gold and black mouldings. The roof boards are of kauri, alternately clear and stained. In each car a clock, set on springs to avoid vibration, is provided. Life guards are fixed to the axles, and the usefulness of these was proved on Saturday evening, when a young man fell off the front platform of one of the cars and would certainly have been run over if the contrivance had not thrown him off the line.

Left: A four-horse team hauling two New Brighton Tramway Company horse cars in Seaview Road. The leading car is No. 10, which became No. 91 in the CTB fleet, and has been restored to its original New Brighton Tramway Company livery by members of the Tramway Historical Society. *W.A. Kennedy collection, Canterbury Museum 11524*

Below: The same No. 10 fully restored to 1894 condition, giving rides on the Ferrymead Tramway at Christchurch in the 1980s. *Graeme Clephane*

Left: Renumbered No. 91 by the Christchurch Tramway Board and saloon windows increased from four to eight. *Alexander Turnbull Library G2752 1/1*

NEW BRIGHTON TRAMWAY COMPANY

The fleet according to 1997 research:

Fleet No.	Builder	Date	Type
1, 2	Moor & Co, Christchurch	1886	Punt or Amercian Summer car.
3, 4	Moor & Co, Christchurch	1887	Double-decker, eight-window saloon.
5	Boon & Stevens Christchurch	1889	Double-decker, saloon, seven-window saloon, built as a single-decker, but designed to have an upper-deck added – originally an 18-seater. Became CTB No. 87.
6	Moor & Co, Christchurch	1889	Double-decker, seven-window saloon, one-horse. Became CTB No. 92.
7	Boon & Stevens, Christchurch	1892	Double-decker, eight window saloon.
8	Moor & Co, Christchurch	1893	Double-decker, eight-window saloon.
9, 10	Boon & Stevens, Christchurch	1894	Double-decker, eight-window saloon.

All the double-decker eight-window saloon cars had four of the window frames covered with solid panels and no end bulkhead windows.

Like the majority of New Zealand steam-tram trailers and horse-drawn trailers, few official records exist.The company had leased from 1888 two Booth Macdonald cars built for the Christchurch City Council line and it is *recorded that the Christchurch Tramway Board took over ten cars from the New Brighton Company on 1 August 1905.

It seems by the time the company sold to the Christchurch Tramway Board in August 1905, the rolling stock was in need of much maintenance as this report in the *Lyttelton Times* stated on 02/06/05:

> As illustrating the bad state of the New Brighton tramcars, the strange spectacle was witnessed yesterday of a traveller seated inside the smoking car with his umbrella open. The rain was coming through the roof in quite a stream and complaints were loud and varied. There is talk of an agitation for a motor-bus service to the borough.

The New Brighton Tramway Company's two "Punts" were not added to the Christchurch Tramway Board fleet and only one of the Christchurch Council saloon double-deckers (No. 2) saw service with the CTB.

*Christchurch Tramway Board fleet numbers assumed to be 86 to 94, one of which was Christchurch City Council car No. 2. A list dated 12 March 1903 of the New Brighton Tramway Company fleet stated "eight large cars, two open cars, exclusive of two cars belonging to the City Council".

* David Hinman research

Above: The city departure point for New Brighton Tramway Company horse-trams was in front of Warner's Hotel in Cathedral Square. *Alexander Turnbull Library F25301½*

Below: Through open country the line ran to the sand dunes at New Brighton. A Kitson steam tram locomotive and saloon double-decker trailer can be seen in the distance at the end of the line. *Alexander Turnbull Library. G358 1/1*

City & Suburban Tramway Company

Horse-Trams

Fleet Numbers: Known numbers 1, 2, 9, 10 (six cars)
Year Built: 1893 – No's 1 & 2 – rest later.
Withdrawn: Taken over by Christchurch Tramway Board 1906.
Builders: The first two cars were built by W. Moor & Co, Christchurch.
Specifications as built: Single-truck, double-deckers, single lower saloon with longitudinal seats, curved lower rocker panel, straight upper waist panel, open upper deck with knife-board centre back-to-back longitudinal seating, left-hand quarter-turn 90° staircases at each end, open fronts.

All the C&STC saloon double-decker horse-cars had four windows only on each side of the saloon and the other four window frames were covered with solid panels, no bulkhead windows.
The company hired one horse-tram for a month from the Canterbury Tramway Company when the service opened in September 1893.

Horse-Trams

Fleet Numbers: Unknown (four cars)
Year Built: 1890s
Withdrawn: Taken over by Christchurch Tramway Board 1906.
Builder: Local – probably Christchurch Tramway Company or Boon & Stevens.
Specifications as built: Single-truck, open lower and upper decks with knife-board centre back-to-back longitudinal seating, decency boards, left-hand straight staircases at each end, open fronts. Known as a *Cage.*

The rolling stock was part of the asset sale to the Christchurch Tramway Board in 1906.
*The saloon double-deckers became No's 95 to 100 in the CTB fleet.
*The open double-deckers became No's 101 to 104 in the CTB fleet. 101, 102 had curved end aprons, 103, 104 had straight end aprons.

*David Hinman research.

Above: This photograph taken at the New Brighton terminus, the corner of the Esplanade and Seaview Road, shows the two types of horse-trams used by the City & Suburban Tramway Company from 1893. The letterboard above the saloon windows reads "Richmond, Burwood & New Brighton". *Tramway Historical Society*

City & Suburban Tramway Company, Ltd.

L'ave Clock Tower	St'nm're Br'dg	Dudley Creek	Buxton's Corner	Burwood	Brooklyn	New Brighton Beach.
am	am	am	am	am	am	am
7,0	7,15	7,20	7,25	7,35	7,45	7,55
8,0	8,15	8,20	8,25	8,35	...	...
	...	...	...	...	...	...
9.0	9,15		...	...	..	...
10,10	10,25	10 30	10,35	10,45	10,55	11,5
11,5	11.20	11,25	11.30	11,40	11,50	12,0
pm	pm	pm	pm	pm	pm	pm
12,5	12,20	12,25	..	...	...	...
*1,5	1,20	1,25	1,30	1,40	1,50	2,0
2,10	2,25	2,30	2,35	2,45	2,55	3,5
3,10	3,23	3,27	3,32	3,40	3,50	4,0
4,0	4,15	4,20	4,25	4,35	4.45	4.55
5,10	5,25	5,30	5.35	5,45	5 55	6, 5
6,10	6,25	6,30	6,35	6,45	6,55	7,5
...	...	...	..		..	...
7,15	7,30	7,35	7,40	7.50	8,0	8,10
9,15	9,30	9,35	9,40	9,50	10,0	10.10
...	...	...	...	...	..	...
‡10,10	10,25	10,30	10,35	10,45	10 50	11,5

Leave New Brighton Beach	Brooklyn	Burwood	Buxton's Corner	Dudley Creek	St'nm're Brd'g	Arrive Clock Tower
am	am	am	am	am	am	am
7,0	7,10	7.20	7,30	7,35	7,40	7,55
...	...	...	...	...	8,15	8,30
8,0	8,10	8,20	8,30	8,35	8,40	8,55
...	...	...	...	...	9,15	9,30
...	...	9,20	9,30	9,35	9,40	9,55
..	..	..	..	..	...	...
11,5	11,15	11,25	11,35	11,40	11,45	12,0
pm	pm	pm	pm	pm	pm	pm
...	...	...	...	12,35	12,40	12,55
1,5	1.15	1,25	1,35	1,40	1,45	2,0
2,10	2.20	2,30	2,40	2,45	2,50	3,5
3,5	3,15	3,25	3,33	3,40	3,45	4,0
4,0	4,10	4,20	4,30	4,35	4,40	4,55
..	..	...	...	...	5,30	5,45
5,5	5,15	5,25	5,35	5,40	5,45	6,0
6,10	6,25	6.35	6,45	6,50	6,55	7,10
7,10	7,25	7,35	7,45	7,50	7,55	8,10
...	...	..	...	...	..	..
9.10	9.20	9.30	9,40	9,45	9,50	10,5
10,10	10,20	10,30	10,40	10,45	10,50	11,5

SUNDAY

L'ave Clock Tower	St'nm're Br'dg	Dudley Creek	Buxton's Corner	Burwood	Brooklyn	New Brighton Beach.	Leave New Brighton Beach	Brooklyn	Burwood	Buxton's Corner	Dudley Creek	St'nm're Brd'g	Arrive Clock Tower
am	am	a.m	am	am	am	am	am	am	am	am	am	am	am
10,0	10,15	10,20	10,25	10.35	10.45	10,55	10,0	10,10	10.20	10,30	10,35	10,40	10,55
11,0	11,15	11,20	11,25	11,35	11 45	11.55	11,0	11,10	11,20	11,30	11,35	11,40	11,55
pm	pm	pm	pm	pm	pm	pm	pm	pm	pm	pm	pm	pm	pm
12.55	1,10	1,15	1,20	1,30	...	...	12,0	12,10	12,20	12,30	12,35	12,40	12,55
2,10	2,25	2,30	2,35	2,45	2,55	3,5	..	...	1,30	1.38	1,43	1,47	2,0
3,0	3,15	3,20	3,25	3,35	3,45	3.55	4,0	4.10	4.20	4,30	4,35	4,40	4,55
5,10	5,25	5.30	5.35	5,45	5.55	6,5	5,10	5,20	5,30	5,40	5,45	5.50	6,5
8.25	8.40	8.45	8.50	9.0	9.10	9.20	8.25	8.35	8.45	8.55	9.0	9,5	9.20

SATURDAY NIGHT

Leave Clock Tower	Stanmore Bridge	Dudley Creek	Buxton's Corner	Burwood	Brooklyn	Brighton Beach
p m	pm	p m	p m	pm	p m	pm
6,10	6,25	6,30	6,35	6,45	6,55	7,5
7,15	7,30	7,35	7,40	7,50	8,0	8,10
8,15	8,30	8,35	8,40	...	...	...
9,15	9,30	9,35	9,40	9.50	10,0	10,10
10,15	10,30	10.35	10,40	10,45	11,0	11.10

Leave Brighton Beach	Brooklyn	Burwood	Buxton's Corner	Dudley Crk	Stn'm'r Br'g	Arrive Clock Tower
6,10	6,25	6,35	6,45	6,50	6,55	7,10
...	...	...	...	...	7,30	7,45
7.10	7,25	7,35	7,45	7,50	7,55	8,10
...	...	...	8,45	8,60	8,55	9,10
9,10	9,20	9,30	9,40	9,45	9.50	10,5
10,10	10,20	10,30	10,40	10,45	10 50	11,5

* Thursdays and Saturdays leaves 1.10 p.m.
‡ Saturdays, 10,15 p.m.
Table NOT in Force on Holidays.
Return Fare 10,10 a m, SIXPENCE.

5

Above: From the January 1900 issue of the Smiths Bradshaw Timetable book. The City & Suburban Tramway Company Limited timetable, showing the departure times to and from the Clock Tower and New Brighton Beach, with arrival times at Stanmore Bridge, Dudley Creek, Buxton's Corner, Burwood and Brooklyn. *Alan Bellamy collection*

City & Suburban Tramway Company

Right: The city terminus for the City & Suburban Tramway Company horse-trams was in Manchester Street by the Queen Victoria Jubilee Clock Tower, erected in 1897 at the intersection of Manchester and High Streets. In this photograph seven C&STC horse-trams can be seen, five saloon double-deckers and two *Cage* double-deckers. *E. Wheeler & Son, Canterbury Museum 886*

Below right: Two C&STC saloon double-deckers by the Clock Tower. A Christchurch Tramway Company Kitson locomotive and trailer in the background. The Clock Tower was moved to Victoria Street in 1930. *H.C. Gore collection, Tramway Historical Society*

Below: In 1905 passengers can be seen climbing the front staircase to the top deck for an open-air journey by a two-horse tram to New Brighton via North Beach. An electric tram passes in the background heading for Papanui. *Head collection, Alexander Turnbull Library 9761 1/1*

City & Suburban Tramway Company

Above: City & Suburban Tramway Company saloon double-decker No. 1 crossing the Stanmore Road Bridge bound for North Beach and New Brighton. *F.T.Series, Trevor Terry collection*

Left: Umbrellas on the top decks protect the complexions of the Victorian ladies as a four-horse team raises the dust at a full trot on the City and Suburban line. The tram is seen heading back through Burwood bound for the city terminal. *John Bettle collection*

Christchurch Tramway Company

This was a new company formed following the liquidation of the Canterbury Tramway Company in 1893, and which took over all the assets of the Canterbury company including all the rolling stock. This included a fleet of 22 tramcars, eight steam-trams and 45 horses.

Below: Kitson steam-tram locomotive No. 6 hauling Christchurch Tramway Company *Cage* trailers 24, 35, 44 and 45.
Lower: Kitson No. 7 turning into High Street with CTC *Cage* trailers 44, 33, 39, and saloon double-decker trailer No. 8. See opposite page for details about these trailers. *Alexander Turnbull Library G73964½*

The Canterbury Tramway Company fleet taken over by the Christchurch Tramway Company was as follows:

Fleet Numbers	
1 to 9	saloon deckers built Stephenson.
10 to 16	saloon deckers built Moor & Sons.
17 to 19	Seven-window deckers built Moor & Sons (originally single-deckers).
20	open toast rack with cross-bench seats built Moor & Sons.
21	open decker built Booth Macdonald.
51	single-deck single-saloon built Stephenson.

See pages 87 to 93.

Christchurch Tramway Company

Steam-Tram Trailers / Horse-Trams

Fleet Numbers: (Probable) 22 to 24, 33 to 35, 38, 39, 41, 42, 44, 45 (twelve cars)
Year Built: 1890s
Withdrawn: Taken over by the Christchurch Tramway Board in 1905.
Builder: Local builders – most by the Christchurch Tramway Company in their own yard, others by Boon & Stevens, Boon & Company.
Specifications as built: Single-truck, open lower and upper decks with knife-board centre back-to-back longitudinal seating, decency boards, left-hand straight staircases at each end, open fronts.

Known as *Cage* trailers.
The fleet was gradually expanded to cope with increased services and lines during the 1890s.The Christchurch Tramway Company found *Cage* trailers were economical to construct and could carry a sizeable load.
From photographs, the majority were built with straight staircases, later rebuilt with left-hand quarter-turn 90°curved staircases, possibly by the company.

Left: Trailer No. 22 of the Christchurch Tramway Company, possibily the first tramcar to be built for or by the company. The photograph was taken in 1901 when troops were departing for the Boer War. *Weekly Press, Canterbury Museum 1737, detail*

Below: An inward-bound steam-drawn tram from Sumner with *Cage* trailer No. 44, one of the last to be built during Christchurch Tramway Company days, in the foreground.

Christchurch Tramway Company
Steam-Tram Trailers / Horse-Trams

Fleet Numbers: (possible) 25 to 28 (four cars)
Year Built: 1896
Withdrawn: Purchased by the CTB in 1905 as part of the company assets, but not added to the CTB fleet.
Builder: Christchurch Tramway Company.
Specifications as built: Single-truck, single deck, centre entrance California. Length: 24 feet.

In the 11/09/96 issue of *The Press*, the paper reported on these cars being built by the Christchurch Tramway Company in their own workshops:

> The Christchurch Tramway Company have recently introduced some new cars on to their lines, which, apart from the fact that they are exceedingly creditable speciments of local manufacture – being built, painted, etc., on the premises of the Company – are also very comfortable and natty looking. Three are already running, and a fourth is in course of construction. The cars are 24 feet long, and are seated with longitudinal and cross seats, leaving an alley way between arranged to seat forty persons. At each end is a canopy, which covers the driver and seats for five. This enables those who desire to smoke to do so without incommoding anyone. The cars are entered on both sides from the centre and are fitted with patent spring doors, the opening of one of which opens both, and they are also self-closing. The seats at each end and the centre doors obviates ladies getting on to the car having to push their way through a number of men, besides which there can be no standing on the platform. As had already been said, these cars hold forty people, while the older style cars, inside and outside only hold forty-four, the new cars being very much lighter and more comfortable. A slide window at each end enables the guard to collect the fares from the occupants of the platform seats without inconveniences. The new cars reflect great credit on the Company, and are likely to become very popular with travellers on the line.

Above: A photograph dated 9 November 1899, showing two of these single-deck tramcars with upper curved framed saloon windows. On the left is No. 43 (formerly No. 51), the 22nd and last car purchased by the Canterbury Tramway Company back in 1887, working for the Christchurch Tramway Company. *John Bettle collection*

Right: This newspaper photograph taken in 1933 of shooters at Lake Ellesmere, shows the remains of a similar saloon arched window tram to the two shown in the photograph above. It is believed to be one of these four single-deck cars built by the Christchurch Tramway Company but scrapped when taken over by the Christchurch Tramway Board. *David Hinman collection*

Christchurch Tramway Company
Steam-Tram Trailer / Horse-Tram

Fleet Numbers: 29, 30, 40, 46, 47 (five cars)
Year Built: mid-1890s
Withdrawn: Taken over by the Christchurch Tramway Board in 1905 – the last series of former horse-trams to be converted and used as electric tram trailers. Renumbered by the Christchurch Tramway Board series No's 105 to 109.
Builder: Unknown, possibly Christchurch Tramway Company.
Specifications as built: Single-truck, open toast rack with cross-bench seats, entry/exit running boards, arched roof, glass end bulkheads, canvas weather blinds, open fronts, known as *Punts*.

Below: Punt No. 40 leaving Cathedral Square with guests for the official opening of the electric tram service in June 1905. *F.B. Lucas, Canterbury Museum 6373*

Lower: Three *Punts* preparing to leave the Square for the Falsgrave Street electric tramsheds on the same day. From examination of the photographs, these three *Punts* were No's 40, 46 and 47. *Alexander Turnbull Library G8452 1/1*

Christchurch Tramway Company
Steam-Tram Trailers / Horse-Trams

Fleet Numbers: (probable) 31, 32, 36, 37. (four cars)
Year Built: mid-1890s
Withdrawn: Taken over by the Christchurch Tramway Board in 1905.
Builder: Boon & Stevens built at least two of these cars.
Specifications as built: Single-truck, double-deckers, single lower saloon with longitudinal seats, curved lower rocker panel, straight upper waist panel, open upper deck with knife-board centre back-to-back longitudinal seating, left-hand quarter-turn 90° staircases at both ends, open fronts. Seating: 50.

In 1897 the company took delivery of two new double-decker cars from Boon & Stevens with seating for 50 passengers. By the end of 1897 the new company had 37 tramcars, Five Palace cars (horse buses), seven steam-tram locomotives and 111 horses. To feed the horses 16,000 bushels of oats per annum were required.

Above & right: The Christchurch Tramway Company had a large fleet of saloon double-decker cars. Three were slightly smaller with only seven saloon windows compared with the majority with eight windows. In the above photograph the second car from the right is a seven-window saloon, while in the photograph at right, the middle car is also a seven-window saloon, and it could be No. 19. No's 17 to 19 started life as single-deckers. see page 89.

Christchurch Tramway Company

Water-Tank Trailers

Fleet Number: Unknown (two cars)
Year Built: 1897
Withdrawn: Taken over by the Christchurch Tramway Board in 1905. Scrapped 1913 – converted into metal trucks.
Builder: Christchurch Tramway Company.
Specifications as built: Single-truck, 1300 gallon wooden tank.

Above: The two 1300 gallon wooden tank-trailer sprinklers in the electric tram era on the Worcester Street siding with a Kitson steam tram locomotive in 1907.

Below: This photograph taken in 1902 shows the Christchurch Tramway Company's tramsheds and offices in the south-east corner of Cathedral Square. The two 1300 gallon wooden tank-trailer sprinklers built by the Christchurch Tramway Company are at the head of a depot road at right. A horse-drawn *Punt* is seen entering the depot while a Kitson steam tram is at the head of a depot road on the left. *Weekly Press, Canterbury Public Library, detail*

Christchurch Tramway Company

The *Lyttelton Times* of 6/05/05 stated that the company had started with 22 cars and when sold to the Christchurch Tramway Board in May 1905, had a fleet of 47.

23 closed saloon double-deck cars
No's 1 to 19, 31, 32, 36, 37
6 single-deck closed cars
No's 20, 25 to 28, 43
5 cross-bench cars
No's 29, 30, 40, 46, 47
13 open double-deck cars
No's 21, 22 to 24, 33 to 35, 38, 39, 41, 42, 44, 45

Plus:
7 steam-tram locomotives
2 sprinkler trailers

*The Christchurch Tramway Board renumbered the saloon double-deckers No's 52 to 70, 84 and 85, the *Cage* double-deckers No's 71 to 83. Two double-deck cars were rebuilt into one double-truck trailer by Boon and Company and given the road No. 51.
All Christchurch Tramway Company double-decker eight-window saloon cars had two window areas covered with a solid panel each side of the car and bulkhead windows.

* David .Hinman research.

Above: A train of four trailers and a Kitson steam-tram locomotive preparing to leave Cathedral Square during the Christchurch Tramway Company era. The third trailer is a seven-window saloon and the fourth a *Cage*.

Christchurch Tramway Company

Above: Ten double-deck *Cage* trailers and two saloon double-deck trailers in Cathedral Square for the 50th anniversary procession of the Canterbury Province in December 1900. The saloon double-decker trailer No. 10, in the foreground, was the first street tramway vehicle to be built in New Zealand for the Canterbury Tramway Company by Moor & Son in 1880. *John Bettle collection*

Above: Car No. 7, imported back in 1879 from John Stephenson of New York by the Canterbury Tramway Company, working on the Sydenham service with a pair of sturdy horses in Christchurch Tramway Company days. *Canterbury Museum*

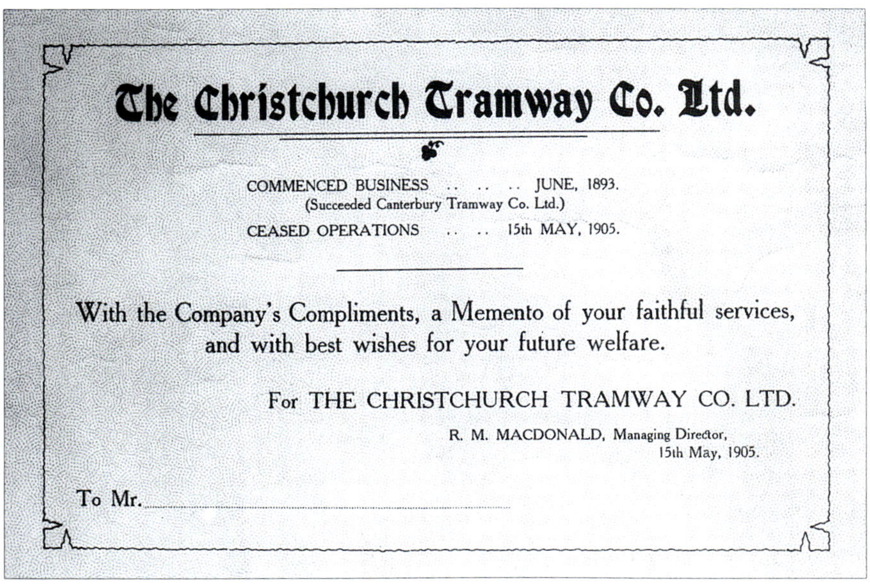

The Christchurch Tramway Co. Ltd.

COMMENCED BUSINESS .. .. .. JUNE, 1893.
(Succeeded Canterbury Tramway Co. Ltd.)
CEASED OPERATIONS .. .. 15th MAY, 1905.

With the Company's Compliments, a Memento of your faithful services, and with best wishes for your future welfare.

For THE CHRISTCHURCH TRAMWAY CO. LTD.
R. M. MACDONALD, Managing Director,
15th May, 1905.

To Mr. ....

Left: The end of an era – the certificate presented to employees when the Christchurch Tramway Company ceased operations on 15 May 1905.

Christchurch Tramway Company

Right: Looking towards the Port Hills from Cathedral Square in the early years of the 20th century. A two-horse tramcar in the foreground is resting before another timetable journey to the suburbs. *Schaef, Wellington*

Right: A Kitson steam tram hauling two saloon double-decker trailers turning into High Street at the start of a run to the seaside suburb of Sumner in 1899. *Hughes*

Right: This postcard boasted the mileage of this Kitson steam-tram locomotive.

Christchurch Tramway Company

Left: A Kitson steam tram with three saloon double-decker trailers crossing the Ferry Road bridge over the Heathcote River on the way to Sumner.

Left: Around the cliffs to Sumner – St Mary's Church picnickers stopped for the photographer by Shag Rock in March 1897. The caption reads "Twenty trams dragged by a single motor". On close examination of the photograph, there are nineteen double-decker trailers and four Kitson steam-tram locomotives.

Below: A Kitson steam tram leaving Sumner in the 1890s with three saloon and two *Cage* double-decker trailers.

Christchurch Tramway Board

Steam Tram

Baldwin Type (One tram locomotive)
Fleet Number: Originally either No. 19 or 23 in the NSWGT fleet. The confusion occurred in the scrapyard from where the locomotive was purchased by the Christchurch Tramway Board in 1906. Given the Fleet number 9 by the CTB.
Year Built: December 1880 or January 1881 for the New South Wales Government Tramways, Sydney, Australia.
Withdrawn: 1923
Builder: Baldwin Locomotive Works, Philadelphia, USA.
Class 4-14-1/3-C9
Specifications as built: Wooden cab. Wheel arrangement: 0-4-2, altered by the New South Wales Government Tramways in 1887 to a 0-4-0. Saddle tank. Wheel base: 4 feet 6 inches. Cylinders: 10 inches x 14 inches. Wheel diameter: 2 feet 11 inches. Working boiler pressure: 150 lbs per square inch. Copper firebox, vacuum brakes, kerosene headlamps.

Purchased second-hand from the New South Wales Government Tramways by the Christchurch Tramway Board in 1906.
Used during the Christchurch Exhibition of 1906 and on other passenger services as horse services were replaced with steam traction haulage before electrification was completed.
*Fitted with NZR couplers and used to shunt coal wagons for the tramway powerhouse on dual-gauge track.

* Alan Bellamy research.

Right: Christchurch Tramway Board No. 9 steam -tram locomotive after being fitted with New Zealand Railways type couplers for the shunting of coal wagons from the NZR siding to the board's siding by the tramway powerhouse.

Below: The Baldwin steam-tram locomotive CTB No. 9 hauling ballast wagons at the intersection of Fitzgerald and Moorhouse Avenues. *Alexander Turnbull Library*

Christchurch Tramway Board

Former Canterbury Tramway Company (CaTC), Christchurch City Council (CCC), Christchurch Tramway Company (ChTC), New Brighton Tramway Tramway Company (NBTC), and the City & Suburban Tramway Company (C&STC), trailers – which were taken over by the Christchurch Tramway Board. Rebuilt and given new fleet numbers by the Christchurch Tramway Board for service as electric tram trailers. Conversion for use as electric tram trailers included the fitting of gates, new wire work, air brakes, electric lights and buzzers, new axles with steel wheels and tyres.

CTB Fleet Number	Private Company Fleet Number	Private Company	Type	Number of Saloon windows	Bulkhead windows	Builder	Date Withdrawn	Comments
50	51 / 43	CaTC/ChTC	CSD/Calif	4	Yes	Stephenson	1954	Preserved
51	31,32,36,37?	ChTC?	8whCDD	16	Yes	Boon & Co.	1937-38	Ex 2 BoonS
52	Unknown	CaTC?	CDD				1944-45	
53	31,32,36,37	ChTC	CDD	8	Yes	Boon & Stevens	1954	
54	1 - 16		CDD	8	Yes		1946	
55	1-19?	CaTC?	CDD				1939-40	
56	1-9?	CaTC	CDD	8	Yes	Stephenson?	1950	
57	1-16	CaTC	CDD	8	Yes		1947	
58	1-16	CaTC	CDD	8	Yes		1919	
59	1-19	CaTC	CDD				1921-22	
60	1-9	CaTC	CDD	8	Yes	Stephenson	1947	
61	1-16?	CaTC	CDD				1946-47	
62	1-19	CaTC	CDD				1915-16	
63	10-16	CaTC	CDD	8	Yes	Moor & Sons	1950	
64	1-9	CaTC	CDD	8	Yes	Stephenson	1950	Body at Ferrymead
65	10-16	CaTC	CDD	8	Yes	Moor & Sons	1950	
66	1-16	CaTC	CDD	8	Yes		1920-21	
67	1-16?	CaTC?	CDD	8?	Yes?		1915-16	
68	1-16?	CaTC?	CDD	8?	Yes?		1938	
69	1-16?	CaTC?	CDD	8?	Yes?		1915-16	
70	1-16?	CaTC?	CDD	8?	Yes?		1946	
71*		ChTC	Cage (Conv)	7			1950	
72*		ChTC	"				1954	
73*	21?	CaTC?	"				1947	Poor design
74*		ChTC	"				1953	Preserved
75*		ChTC	" (Conv)	7			1946	
76*	*21, 22-24, 33,	ChTC	"				1946-47	
77*	34, 35, 38, 39,	ChTC	" (Conv)	7			1950	
78*	41, 42, 44, 45.	ChTC	" (Conv)	7			1952-53	
79*		ChTC	" (Conv)	7			1952-53	
80*		ChTC	"				1946-47	
81*		ChTC	"				1946-47	
82*		ChTC	"				1940-41	
83*		ChTC	"				1945-46	
84	17-19	CaTC	CDD	7	No		1947	Body at Ferrymead
85	1-19		CDD				1947	
86	3, 4, 8, 9.	NBTC	CDD	8			1945-46	
87	6•	NBTC	CDD	7	No	Boon & Stevens	1950	
88	3, 4, 8.	NBTC	CDD	8	No	Moor & Co	1945-46	
89	3, 4, 8, 9.	NBTC	CDD	8	No		1947	
90	3, 4, 8.	NBTC	CDD	8	No	Moor & Co?	1952	
91	10•	NBTC	CDD	8	No	Boon & Stevens	1950	Preserved
92	5•	NBTC	CDD	7	No	Moor & Co	1947	
93	2?	CCC	CDD	8		Booth Macd	1937-38	
94	7	NBTC	CDD	8	No	Boon & Stevens	1952	
95*		C&STC	CDD	8	No		1945-46	
96*		C&STC	CDD	8	No	Moor & Co	1947	
97*	*1-10 (some of,	C&STC	CDD	8	No	Boon & Stevens	1947	
98*	including 1-2)	C&STC	CDD	8	No		1946	
99*		C&STC	CDD	8	No	Boon & Stevens	1945-46	
100*		C&STC	CDD	8	No		1954	

CTB Fleet Number	Private Company Fleet Number	Private Company	Type	Number of Saloon windows	Bulkhead windows	Builder	Date Withdrawn	Comments
101*		C&STC	Cage				1952	
102*	*3-10 (some of)	C&STC	"				1945-46	
103*		C&STC	"				1954	- Bike trailer from 1946
104*		C&STC	"				1945-46	
105**		ChTC	Punt				1937-38	
106**	**29, 30, 40, 46, 47.	ChTC	Punt				1937-38	
107**		ChTC	Punt				1937-38	
108**		ChTC	Punt				1937-38	
109**		ChTC	Punt				1937-38	

Notes:

1. Private company numbers where marked with • are proven "beyond reasonable doubt".
2. Where several numbers, or range given, the tram was one of those numbers.
3. ? – Possibly correct – based on information so far obtained.
4. Private Company name is the original company the tram was built for.
5. CDD = closed double-decker (closed single saloon below, open upper deck).
6. CSD = closed single-decker (closed single saloon).
7. Cage = open double-decker (open below and on upper deck).
8. (Conv) = rebuilt from a Cage to a closed saloon.
9. Punt = single-decker with open cross bench seating.
10. Number of side windows on each side of saloons in CTB era (NBTC and C&STC each had four blanked off on each side, and the ChTC had two blanked off on each side.
11. Bulkhead windows significant for identification as it appears that only CaTC and some ChTC-built cars had them. Known to have been filled-in in case of 17-19 CaTC when converted from CSDs to CDDs.

Right: Two single-truck, double-deckers, possibly built by the Christchurch Tramway Company, were rebuilt by the Christchurch Tramway Board and equipped with double-trucks as an experiment for service with electric trams. See page 137 *Always a Tram in Sight.*

Below: CTB No. 101, former straight staircase *Cage* of the City & Suburban Tramway Company and CTB No. 65, a Moor & Sons car built for the Canterbury Tramway Company in the 1880s. *Graham Stewart*

Right: No. 84 after being withdrawn from service in 1947. Originally built as series No's 17 to 19, part of an order of seven-window saloon single-deckers for the Canterbury Tramway Company in 1882 by W. Moor & Sons, Christchurch. Car body at Ferrymead awaiting restoration. *Alan Bellamy*

Former 19th century private company trailers were still being used in the 1940s to take racegoers to the Addington Racecourse. These photographs were taken on the last day of the 1940s – 31 December 1949. *Above:* CTB No. 100 (left), a former City & Suburban Tramway Company horse-tram, and CTB No. 79 (right), a former Christchurch Tramway Company *Cage. Graham Stewart*

Below: CTB No. 64 (left), one of the original Stephenson-built cars for the Canterbury Tramway Company in 1879 and No. 71 (middle), a former *Cage.* Both No's 71 and 79 were rebuilt by the CTB with lower saloons in 1923. In the distance (at right) are trailers 90 and 91, formerly of the New Brighton Tramway Company. *Graham Stewart*

Right: CTB No. 91, formerly New Brighton Tramway Company No. 10, and CTB No. 74, formerly a Christchurch Tramway Company *Cage* trailer. Photograph taken at CTB yard in 1954. Both trailers were rescued from private owners after being sold and have been restored by the Tramway Historical Society of Christchurch. Now working exhibits at Ferrymead, Christchurch. *Graham Stewart*

DUNEDIN

Steam & Horse Tramway

(Steam operation ceased March 1884)

David Proudfoot

From July 1879 to February 1883

Dunedin City & Suburban Tramway Company Limited

From February 1883 to March 1901

Dunedin City Corporation Tramways

From 21 March 1901 to 1904 / 1905

Gauge 3 feet 6 inches
The population of Dunedin was 24,372 when the system opened.

The Suburbs and Streets served by the Steam and Horse-Trams

From Cargill's Monument to the Water of Leith via Lower High Street, Castle Street 7 July 1879.
To the Botanical Gardens via Princes, George, Albany and Great King Streets, opened shortly after 7 July 1879.
Branch from the Monument down Rattray Street to car barn and Rattray Street Wharf. Physical connection with New Zealand Railways at wharf approach. Wharf tracks used by both operators. July 1879.
Gardens to Lindsay's Bridge 3 October 1879; to Kelvin Grove 1 January 1880; to Normanby September 1881; to Hawthorne (Bethunes Gully) 1884.
Princes Street to The Oval 3 November 1879; Andersons Bay Road to the Caledonian Sports Ground 1 January 1880. Branch line in Andersons Bay Road to the Caledonian Grounds closed 1882.
Princes Street South, Cargill's Corner, along King Edward Street and Prince Albert Road to St Kilda 21 February 1880; also branch line along Richardson Street to original Forbury Park Racecourse, near Hargest Crescent 1880.
Branch from Cargill's Corner along Hillside Road into David Street, via Forbury Corner, to College Street, Caversham 25 May 1880.
From Forbury Corner (cnr David Street and Hillside Road), along Forbury Road to St Clair 10 December 1883.

David Proudfoot

David Proudfoot

Steam-Trams

Hughes Type (Four tram locomotives)
Fleet Names: *City of Dunedin, New Zealander, Empress, Duchess* (the name *Duchess* never confirmed)
Order of names in relation to maker's numbers unknown.
Year built: 1878
Withdrawn: One engine withdrawn after depot fire in December 1880, possibily for spare parts.
Remainder withdrawn in March 1884 and stored at the Kelvin Grove depot, North East Valley.
Builder: Hughes Locomotive & Tramway Engine Works, Loughborough, England. Maker's numbers 326 to 329.Maker's No. 329 became the fleet No. 4 following the 1880 fire.
Specifications as built: Wheel arrangement: 0-4-0. 7 inch x 12 inch outside cylinders. Wheel diameter: 2 feet 6 inches. Boiler diameter: 2 feet 4 inches. Saddle tank. Boiler pressure: 140 lbs per square inch. Weight in working order: 7 tons. Maximum service speed: 15 miles per hour. City by-laws limited speed to 10 miles-per-hour.

Coke-fired to keep smoke to a minimum. The condensers were designed to consume the exhaust steam, through water cooling pipes on the sides of the cab. The condensate and cooled water was drained into tanks placed at each end of the locomotive frame, which was not re-circulated, but discharged into a sump at the Monument, while fresh water was being taken aboard. These were the only Hughes tramway locomotives built with surface condensers. Previous models had cold-water jet condensers. Soon afterwards (in 1880) Hughes lost control of his company and production of Hughes tramway locomotives ceased.
City of Dunedin used for the first trial runs on 12 June 1879. The same steam-tram used to hauled the first public tram from the Exchange to the Water of Leith via Castle Street on 7 July 1879.
Condensers disconnected in 1881 to increase the hauling power.
The three remaining locomotives sold to the Dunedin City and Suburban Tramway Company Limited in February 1883. Steam tramway operation ceased March 1884.
No. 326 was sold by DCC in 1902 and was in bush tramway service in the Tuatapere district until 1920.
No. 327 was sold by DCC in 1902 and was used to drive factory machinery in Dunedin. In 1904 was converted into a bush hauler and worked near Greymouth until 1915.
No. 328 – no trace recorded of this locomotive after December 1880 fire – used for spare parts?
No. 329 converted to winding engine at Shag Point Colliery 1890.

This photograph, taken when the tramways celebrated their first birthday on 7 July 1880, of a Hughes tram locomotive shows clearly the cooling pipes on the side of the cab. The gentleman in the top hat on the left of the locomotive is Mr David Proodfoot. His engineer, a Mr Craig, is leaning against the front of the tram engine.

A lithograph drawing of Dunedin steam-trams outside the old Exchange building in Princes Street which was published in the *Australian Sketcher*. The drawing depicts a Hughes tram engine on the left with the wording *Kelvin Grove* written on the cab, and the Baldwin tram engine named *Atlantic* hauling two trailers on the right. *Ron Grant collection*

This photograph of one of the Hughes tram engines is said to have been taken after the December 1880 fire, the twisted roof having been caused by the heat from the depot fire. A taller funnel had been fitted and the back platform is stacked with bags of coke.

Steaming along Princes Street with a trailer in tow is another Hughes tram engine with a slightly deeper letter board above the cab windows. *Burton Brothers.*

Baldwin Type (One tram locomotive)
Fleet Name: *Atlantic* – renamed *Gordon* after December 1880 depot fire.
Fleet Number: No. 1 following the 1880 fire
Year Built: 1879 – Entered service 14/9/79
Withdrawn: March 1884
Builder: Baldwin Locomotive Works, Philadelphia, USA. Maker's number 4522 – class: 4-13-C9.
Specifications as built: Wheel arrangement 0-4-0. 9 inch x 12 inch horizontal cylinders. Wheel base: 6 feet 6 inches. Wheel diameter: 31 inches. Vertical boiler diameter: 34 inches. Firebox diameter: 28½ inches. Heating surface: 137 square feet. Tank: 150 gallons. Wooden cab. Weight in working order: 7.2 tons. No condenser. 60 per cent more powerful than the Hughes locomotives. No condensers were fitted to control the escape of smoke and steam.
Cab rebuilt after depot fire and resumed service in April 1881 renamed *Gordon*. In later years, because of the clanking of the side rods and the piercing sound of the loud steam whistle, was given the nickname *Howling Gordon*.
Sold to the Dunedin City and Suburban Tramway Company Limited in February 1883. Retained the fleet number 1 by the DC&ST Company.
Steam tramway operation ceased March 1884 by order of the DCC. *Gordon* put up for auction November 1901; no buyers. Boiler used for depot steaming.

Atlantic decorated with ferns and flowers on 7 July 1880 for the first birthday celebrations in the Octagon. The sign hanging from the centre of the cab announces "Atlantic Forever".

The *Atlantic* waiting for departure time at the Princes and High Street terminus. In the background are horse-trams that served the southern suburbs. *Alexander Turnbull Library 2366½*

Baldwin Type locomotive – Brill coachwork
(One combined engine and passenger car)

Fleet Name: *Washington*
Year Built: 1877. Entered service 19/9/79
Withdrawn: 24/10/79 after a fatal accident and rebuilt.
Locomotive Builder: Baldwin Locomotive Works, Philadelphia, USA. Maker's number 4139. Class: 4-9-C6.
Coachwork Builder: J.G.Brill and Company, Philadelphia, USA.
Specifications as built: Wheel arrangement: 0-4-0.
Combined single-ended steam streetcar, passenger entrance from a rear platform. Saloon for 20 seated passengers. 8 inch x 10 inch horizontal outside cylinders. Wheel diameter: 31 inches. Wheel base: 6 feet. Vertical boiler diameter: 30 inches. Firebox diameter: 24-7/8 inches. Length: 23 feet 8 inches. Tank Capacity: 100 gallons. Weight in working order: 9.4 tons. No condenser.

Originally built for the 3-foot gauge Galveston, Brazos & Colorado Railroad which went into receivership. Baldwin repossessed the locomotives and resold them, *Washington* being regauged to 3 foot 6 inches and sold to Proudfoot.
Rebuilt as a steam-tram locomotive with no passenger accommodation March 1880 and renamed *Grand Pacific* after the Grand Pacific Hotel at St Kilda which was owned by Proudfoot. The boiler was placed in the centre of a shortened main frame, a new locally built cab similar in design to the Baldwin wooden steam-tram cabs was fitted with a clerestory coach-style roof.
Apparently not used after the depot fire of December 1880, but was still listed on the rolling-stock roster when the operations were sold to the Dunedin City and Suburban Tramway Company Limited in February 1883.
Sold to Hungerford & McKay, contractors, Greymouth harbour 1886 and used as a shunting locomotive. Between 1890 and 1910, boiler used for a bush hauler at Paroa.

The coach body was converted into a trailer bearing the original name *Washington* on the upper side saloon panels. According to research by R.D. Grant and J.A. Dangerfield,* the coach body was mounted on a leaf-sprung, railway-type underframe, similar to Scandia car No. 30. No. 30 had been damaged by racegoers in February 1880 and the Scandia body may have been replaced with the *Washington* car body.

* *Steam and Horse Tram Working in Dunedin 1879 – 1905* R. D. Grant and J.A. Dangerfield 1991

Above: A publicity photograph supplied by the Baldwin Locomotive Works of their "Steam car for street railroads". It is understood *Washington* was designed to the same pattern as the vehicle featured in this photograph, except the clerestory roof was square at each end. *Ron Grant collection*

Left: A lithograph illustration from the company's catalogue showing the frame, driving equipment and boiler before the coach body is fitted to a combined engine and passenger car.

Above: Washington after being rebuilt as a steam-tram locomotive with no passenger accommodation in March 1880 and renamed *Grand Pacific*. The decorations of flags and greenery were for the first anniversary of the tramways on 7 July 1880. The trailer was a *Hoboken* type built by John Stephenson Company Limited of New York, known in Dunedin as the **New York* cars. They were similar to the larger cars being used by the Wellington Tramway Company. *Ron Grant, Jim Dangerfield research

Below: The coach body of *Washington* after conversion into a trailer, still bearing the original name. This photograph was taken on the same day as the photograph above, when Mr Proudfoot celebrated the first year of tramway operation in Dunedin. The other trailer, No. 32, is thought to have been built by John Stephenson of New York.

Kitson Type (One tram locomotive) No. 1 standard type.
Fleet Name: Not known
Year Built: August 1879
Entered service 15/1/80 on a test run to Kelvin Grove, North East Valley.
Withdrawn: Badly damaged in the depot fire of December 1880. Remains sold to Conyers and Davidson, of the Otago Foundry and rebuilt as a bush tramway locomotive. In 1882 sold to NZ Pine Company, Woodend, Southland, fitted with saddle tank, new cylinders and trailing truck, worked in various locations until 1940.
Builder: Kitson and Company, Airedale Foundry, Leeds, England.
Maker's number: T7
Specifications as built: Wheel arrangement 0-4-0. 6¼ inches x 12 inches stroke cylinders. Wheel diameter: 2 feet 3 inches. Boiler barrel: 4 feet 3 inches long x 2 feet 5 inches in diameter. Firebox: 2 feet 6 inches long x 2 feet 10 inches in diameter. Tubes: 72 tubes 1½ inches in diameter. Heating surface: tubes 96.6 square feet; firebox, 19.4 square feet; total 116 square feet. Grate area: 5.17 square feet. Weight in working order: 5¾ tons. Rating: 20 horsepower. Roof-mounted surface condenser, coke fired.

This Kitson locomotive, although not as powerful as the Baldwin tram locomotives, was acknowledged by Proudfoot's tramways engineer as the best suited to Dunedin conditions. This was the prototype Kitson standard tram locomotive of which at least 267 examples were built for tramways world-wide between 1879 and 1900.

Kitson Type (One tram locomotive) No. 1 standard type.
Fleet Name: Not known – probably given a fleet number.
Year Built: 1882
Entered service late 1882 – to replace the original Kitson tram locomotive withdrawn after the December 1880 depot fire.
Withdrawn: March 1884 and stored at the Kelvin Grove depot, North East Valley.
Builder: Kitson and Company, Airedale Foundry, Leeds, England.
Maker's number: either T52 or T53
Specifications as built: Wheel arrangement 0-4-0. 6¼ inches x 12 inches stroke cylinders. Wheel diameter: 2 feet 3 inches. Boiler barrel: 4 feet 3 inches long x 2 feet 5 inches in diameter. Firebox: 2 feet 6 inches long x 2 feet 10 inches in diameter. Tubes: 72 tubes 1½ inches in diameter. Heating suface: tubes 96.6 square feet; firebox, 19.4 square feet; total 116 square feet. Grate area: 5.17 square feet. Weight in working order: 5¾ tons. Rating: 20 horsepower. Roof-mounted surface condenser, coke fired.

Sold to Gibson & Lee's Burnwell Colliery, South Otago, which had a railway connection to Lovells Flat station yard, 1896. Resold to Komata Reefs Gold & Mining Company in 1899. From 1905 was a bush tramway locomotive in North Auckland.

An early maker's drawing of the Kitson No. 1 standard engine, similar to those supplied to Christchurch and Dunedin. The longitudinal arrangement of condenser tubes was superseded before construction. *Ron Grant collection*

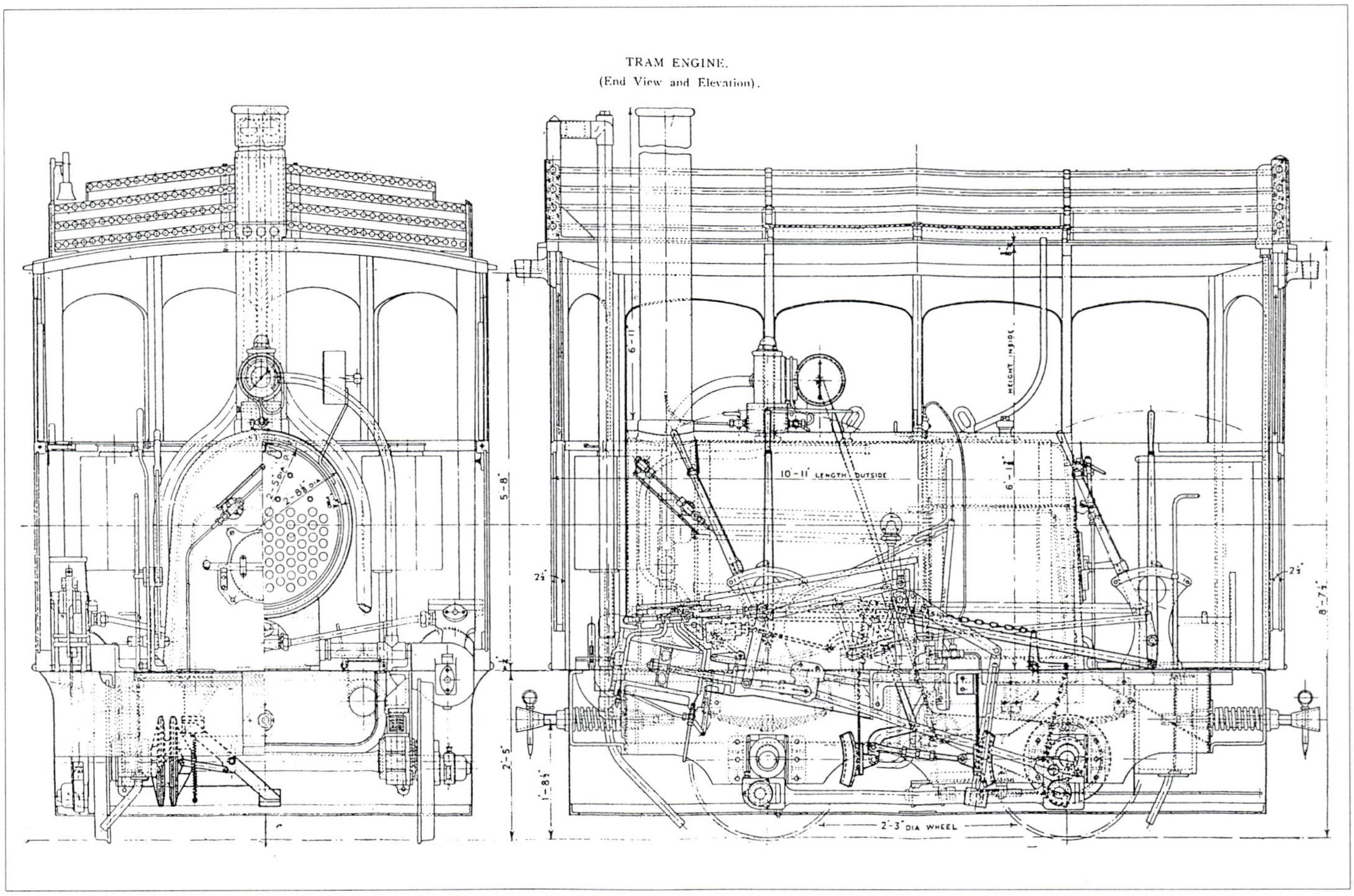

Above: This lithograph of Princes Street, Dunedin, showing three steam-trams, was published in the *Illustrated New Zealand Herald* on 16 June 1881. The tram locomotive in the foreground appears to be a Kitson. *Ron Grant collection*

Left: The second Dunedin Kitson tram locomotive working for the Naumai Sawmills in Northland in 1911.

Horse-trams / Steam-Tram Trailers

David Proudfoot, a civil engineer and land developer, who was granted a tramway concession for 14 years at a rental of £360 per annum by the Dunedin City Council in July 1877, was a forward-thinking entrepreneur. He had instructed his Dunedin agent to obtain the best available equipment from Europe and America. The agent, Mr Hugh Macneil of Dunedin, during his travels visited the Paris Exhibition of 1878 and ordered four steam-tram locomotives, the prototype having won the Exhibition Silver Medal. These locomotives, built by the English company of Hughes Locomotive & Tramway Engine Works, as the previous pages detail, were part of the original steam tram order for Dunedin, together with two Baldwins and one Kitson.

Macneil ordered the two Baldwins when he visited the States and Kitson's Australian agent had visited Proudfoot in Dunedin. Proudfoot, unlike other tramway operators of the period, did not place his faith in one manufacturer.

Proudfoot adopted the same policy in the ordering of horse-trams/steam-tram trailers. Although the majority were originally ordered from American coach builders because of their quality reputation in the 1870s, a variety of cars were obtained from Europe and an order was also placed locally. It was Proudfoot's intention to test a range of cars and then use local coach builders to build copies – the pirating of designs and patents seemed to have been a common practice in those years.

Cars were ordered and shipped from many companies including John Stephenson of New York, J.G. Brill of Philadelphia, Scandia Company of Denmark, Hughes Locomotive & Tramway Engine Works of Loughborough, England, and the Starbuck Company of Birmingham, England. Sinclair and Moreton, corner of St. Andrew and Great King Streets, of Dunedin, and Guthrie and Larnach of the New Zealand Woodware Factory of Dunedin received their first order for six cars.

The *Otago Witness* of 28 June 1879 reported the fleet would comprise seven engines, 32 imported cars and six being built by the New Zealand Woodware Factory.

In October 1879 the *Saturday Advertiser* reported that the tramcars obtained for the new tramway of David Proudfoot were as follows:

> Fifteen 25-seat passenger cars by John Stephenson & Co, New York. Two 52-seat passenger cars by John Stephenson & Co, New York. Three 30-seat passenger cars by John Stephenson & Co, New York. Two 20-seat passenger cars by John Stephenson & Co, New York. One 30-seat passenger car by the Starbuck Company, Birkenhead, England. Two 60-seat passenger cars by Scandia, Norway. Six 25-seat passenger cars by Guthrie and Larnach, Princes Street, Dunedin. One 25-seat pasenger car by Sinclair and Moreton, Dunedin.
> Locomotives:
> Four made by Hughes & Company, Loughborough.
> One made by Baldwin & Company, Philadelphia.
> One made by Kitson & Company, Leeds.
> Horse cars – 60 horses and one, two and three-horse cars.
> Lines open so far were: Double line: Princes, George, Great King Streets to Botanical Gardens and extensions to Lindsays Creek, North East Valley. Single line: Cargill's Monument, Castle Street to Water of Leith. Single line: Rattray Street Steamer Basin, Car and Engine shop. Rails 42 lbs per yard grooved Bessemer steel in 27 feet lengths.

As there had been no official opening ceremony, a celebration was held on the first anniversary of the tramways with the trams decorated with foliage, ferns and flags. All trams stopped running at 7 pm to enable the staff to attend a grand anniversary ball and supper. The fleet by then totalled seven steam trams and 38 cars.

For the complete story, *Steam and Horse Tram Working in Dunedin 1879 – 1905* by R.D.Grant & J.A.Dangerfield, ISBN 0-473-01301-0 published in 1991.

The original horse-tram / steam-tram trailer cars of 1879, built by the Hughes Locomotive and Tramway Works, of England, which were found to be hard riding. The window fittings also gave trouble and after the American cars arrived, they were not used in regular service. This view looking along Moray Place towards the First Church of Otago on Bell Hill, shows on the left, the headquarters of the Burton Brothers, the famous 19th century photographers. Proudfoot had purchased the second First Presbyterian Church, a large wooden building in Lower Dowling Street and shifted it to Willis Street to use as his tramway depot. *Burton Brothers*

Right: Some of the original cars imported by David Proudfoot included No. 16 with solid wheels. This photograph was taken during the first anniversary celebrations in the Octagon. The coach builder remains a mystery. For some of the original John Stephenson Company of New York cars imported by Proudfoot, see pages 134 and 135.

Left: No. 26 of the Proudfoot fleet was built by the Starbuck Car and Wagon Company of Birkenhead, England. *Science Museum / Science & Society Picture Library, London*

Below left: Two 50-seater, double-decker cars ordered by the NZR from the Scandia Company of Denmark for the Wellington Hutt Valley service at the same time as Proudfoot had ordered two cars from the same builders, were shipped to Proudfoot when the NZR cancelled their order. The NZR cars, although heavier than conventional horse-tram design, were suitable for steam-tram operation. No. 29 had a teak body and india rubber springing. The clerestory roof windows under the upstairs longitudinal seating can be clearly seen in this photograph. *Ron Grant collection*

The second Danish built car by Scandia, No. 30, which was originally being built for the Wellington-Hutt Valley service, photographed before being shipped to New Zealand in August 1879. The car was sheathed in iron panelling and had conventional railway leaf springs. The car was destroyed in the December 1880 depot fire. The Scandia-built cars became known as the **Danish Palace Cars* because of their stylish finish. *Ron Grant collection*

*Ron Grant, Jim Dangerfield research.

Right: Front-on views of the two Scandia-built cars. No. 30 was equipped with saloon window curtains. *Ron Grant collection*

Left: During the first year of the tramways, six cars were built locally, five by Guthrie and Larnach of the New Zealand Woodware Company, and one by Messrs Sinclair and Moreton. This photograph of No. 11, a Dunedin-built car, shows an arch roof which extends over the end platforms, compared with the separate canopies on the other locally built cars as illustrated on the opposite page.

Above: A three-horse team hauling No. 9, a Dunedin-built single-deck four-wheel tram by the South Road, King Edward Street intersection. *Tramway Historical Society*

Left: The same tram, No. 9, with a two-horse team at the Manse Street terminus in Princes Street. It is thought that the cars built to this design before the depot fire of December 1880 and destroyed, were rebuilt to the same design and given the same road numbers. *Otago Settlers Museum*

Dunedin coach builders constructing a car to a slightly different design. The car has no fanlight windows above the main saloon windows and the arched roof has a small clerestory for ventilation in the enclosed saloon. *Ron Grant collection*

At the top of the rise out of the Exchange in Princes Street, behind the Leith tram can be seen a "hill boy" on a heavy draught horse which had just assisted the tram up the incline. He will walk his horse down the hill and wait for the next tram.

At 4 am on Sunday, 4 December 1880, the wooden tram depot in Willis Street near the Rattray Street Wharf, was destroyed by fire. Stored on five depot roads were seven steam trams, 36 trailer cars and one goods wagon. Only two trailer cars were saved from the flames.Together with two trailer cars and one goods wagon at the Ocean Beach Depot, they were all that remained of the fleet. As all the company records were destroyed in the blaze, transport historians ever since have been puzzled when studying photographs of the rolling stock as to the builders of such a diverse range of vehicles supplied by so many coach-building companies. Proudfoot lost no time in resuming services with the constructon of 20 temporary cars being started within three days using salvaged equipment. These trailer/horse cars were nicknamed "Skeleton cars" as they were just a basic frame with six cross-bench seats and no roof.

Proudfoot immediately placed orders with coach builders to replace the cars destroyed: six cars from John Stephenson and Company, four from J.G. Brill, and four from Brussels. In June 1881 six saloon cars which had been ordered before the fire arrived from the J.W. Jones Company, Troy, USA. Guthrie & Larnach completed the sixth car of their 1879 contract and were awarded a second contract to build a further five cars. Some of the replacement cars were built to the same design as the first generation cars and given the same fleet numbers as the originals, adding more confusion to the work of historical researchers in this century.

The *Skeleton* cars were withdrawn from daily service in the following September and stored. 24 new cars, three refurbished steam trams and 287 horses were soon running the services while nine cars were still to be delivered from the builders.

Above: No. 15 in the foreground is a *Skeleton* car which has had a roof added. Photographed at the Manse Street terminus in Princes Street. *Ron Grant collection*

Four locally built railway passenger carriages of light construction were lent by the Government railways and used on the Ocean Beach service after the fire of December 1880. *Above:* An NZR *Hislop* car in temporary service for Proudfoot in Princes Street by Stafford Street. *Hardwicke Knight Right:* Another Hislop car heading for Ocean Beach.

Above: A Jones Car Company of America *Governor Strong* design car photographed outside the Grand Hotel in Princes Street. The lantern top above the clerestory roof with the words "Dunedin", was coloured glass which was illuminated at night from the central interior dome lamp. The *Otago Daily Times* described these cars as: "although not extraordinarily large, they are most handsomely finished, the exterior being most artistically decorated in a Scotch plaid design." The seats were perforated wood with cushions, and Venetian blinds were fitted to the windows. *Hocken Library*

Above: No. 54 was one of two single-deck cars, ordered after the fire from J.G. Brill and Company, Philadelphia. The car is standing at the St Clair terminus in 1888, although the destination board hanging from the centre of the rocker panel reads: "To & From Manse St & Caversham". *David Todd*

Two double-decker cars were also ordered from J.G. Brill and Company at the same time as the two single-deckers in 1880. They were distinctive by the arched saloon windows and only had seven windows each side of the saloon compared with the Stephenson cars with eight saloon windows.

Above: One of the Brill double-deckers with a leader horse attached for the pull up the Princes Street grade to the Octagon. *David Todd*

Right: A Brill double-decker in the suburbs with a three-horse team. The driver was George Hill. *Russell Grigg collection*

John Stephenson Company Limited of New York built three double-decker cars for Proudfoot of Dunedin.
Upper: No. 201 in Princes Street. *Field Hodgkins collection, Alexander Turnbull Library, detail 105941½*
Above: No. 202 before modesty boards were fitted. *Burton Brothers*
Below and right: Stephenson double-deckers working the St Clair service out in the country.

In February 1883 Proudfoot sold the tramways to the Dunedin City and Suburban Tramway Company Limited. The rolling stock comprised six steam trams, 32 horse/trailer cars and 307 horses.

For the Dunedin & South Seas Exhibition of 1889-90, the 1880 *Skeleton* cars were used to handle the increased traffic; they had been re-roofed and some fitted with windows.

After much negotiation the Dunedin City & Suburban Tramway Company was sold to the Dunedin City Corporation, the take-over date being 21 March 1901. On 14 February, five weeks before the council took control, fire destroyed 16 horse-trams at the Cumberland Street depot, leaving barely enough cars to run the services until the system could be rebuilt for electric traction. Once again the *Skeleton* cars were placed back in service.

Above: Single-deck American-built cars by John Stephenson, No's 103 and 104. *Alexander Turnbull Library, detail F 20427½ and Field Hodgkins collection, Alexander Turnbull Library F105946½*

Above: No. 109 on delivery from the coach builders in the 1880s to the Dunedin City & Suburban Tramway Company, at the Cumberland Street depot. The builder is unknown. *Otago Settlers Museum*

Above and right: This Dunedin ten-window saloon horse-tram had had a local body graft – two small five-window 10-foot body saloon trams built by John Stephenson of New York had been joined together to make this rather long-looking tram with only four-wheels. *Otago Settlers Museum*

Dunedin City & Suburban Tramway Company Limited

Above: A busy horse-tram scene in Princes Street by the intersection with High Street in the 1880s. *Alexander Turnbull Library F105942½, detail*

Left: From the same intersection with Princes Street, looking up High Street to Mornington. When horse-trams and cable trams waited for the other's commuters. *Otago Settlers Museum.*

Dunedin City & Suburban Tramway Company Limited

Right: A two-horse tram advertises the new penny section fare on the roof canopy. The photograph was taken at the St Kilda terminus outside the Ocean View Hotel, c1892. *Otago Settlers Museum*

Right: A Water of Leith horse-tram passing through the Octagon. *Otago Settlers Museum*

Below: Horse-trams in Princes Street by the High Street intersection during the visit of the Duke and Duchess of Cornwall and York in 1901. The horse-tram on the right is a *Skeleton* car.

A representative of this class has been restored by the Tramway Historical Society

Left: Three horses prepare to take the grade out of the Exchange area in Princes Street. *Trevor Terry*

Below left: Surrounded by the competitors of the period, hansom cabs wait for patrons alongside the horse-tram. *Tramway Historical Society*

Below: The body of No. 18 arrives by rail from Dunedin for restoration by the Tramway Historical Society of Christchurch. *Colin Loach*

Above: No. 18 re-entered service after restoration for use as an electric tram trailer when the Christchurch tourist tramway service opened on 4 February 1995. This November 1995 photograph was taken when the car was hired and used as a horse-drawn tram for a wedding. *David Hinman*

The city celebrated the purchase of the tramway system from the City and Suburban Tramway Company with a parade through the centre of the city. Councillors and guests rode on the trams in the procession and this was followed by free rides for the citizens. The four trams visible in the photograph are from left, a Stephenson double-decker with the sign along the top deck "Advance Dunedin", a *Skeleton* car, a Brill double-decker, and another *Skeleton* car. *Russell Grigg collection*

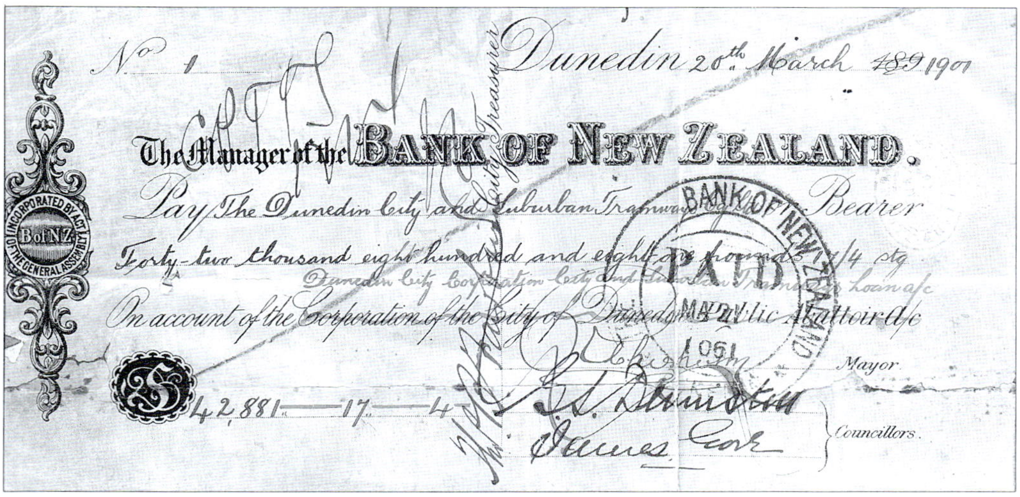

No 1 Dunedin 20th March 1901

The Manager of the BANK OF NEW ZEALAND.

Pay The Dunedin City and Suburban Tramway ... Bearer

Forty-two thousand eight hundred and eighty ... pounds 17/4 stg

Dunedin City Corporation City and Suburban Tramway Loan a/c

On account of the Corporation of the City of Dunedin Public Abattoir a/c

BANK OF NEW ZEALAND PAID

Mayor

£42,881—17—4

Councillors.

The Bank of New Zealand cheque dated 20th March 1901, made out to the Dunedin City and Suburban Tramway Coy Ltd for the sum of £42,881.17.4, on account of the Dunedin City Corporation and Suburban Tramway Loan a/c and signed by the mayor and two councillors. The price of the sale of the horse-tramway system to the city would have been 7.5 million dollars on today's values. *David Todd collection*

Below: The suburban depot at St Kilda in the final days of the Dunedin horse-trams. *Russell Grigg*

DUNEDIN

Steam Tramway

(Operated as a Steam Tramway from 17 January 1881 to 31 May 1882)

The Dunedin, Peninsula & Ocean Beach Railway Company Limited

From 26 October 1876 to 5 January 1881
(In liquidation 5 May 1883)

Dunedin, Forbury & Peninsula Railway & Tramway Company

(Incorporated 11 February 1884, mortgage foreclosed 16 September 1887)

Dunedin City & Suburban Tramway Company Limited

21 April 1888 to 16 March 1901

Dunedin City Corporation

16 March 1901
(Line closed November 1938)

Gauge 3 feet 6 inches
The population of St Kilda was 773 and South Dunedin 2796 when the system opened.

The Suburbs and Streets served by the Steam Trams

THE DUNEDIN, PENINSULA & OCEAN BEACH RAILWAY COMPANY LIMITED
City to Forbury Park: From the Railway Station (Queen's Gardens site), via Railways Department line to its junction with the company line near Kensington; then via line along Upper Harbour foreshore on the northern side of Andersons Bay Road (with spur line to City Gasworks), to Musselburgh and Ocean Beach entrance, to Forbury Park via Royal Crescent and Victoria Road. Line shortened c1899.
7 January 1878: Branch from Musselburgh (reverse junction), via Portobello Road, along the foot of Sunshine Hill (now Shore Street); crossed the Andersons Bay Inlet on embankment to station at the foot of Silverton Street.
DUNEDIN, FORBURY & PENINSULA RAILWAY & TRAMWAY COMPANY
From Rattray Street via Crawford Street (street running), on own tracks to the northern side of Andersons Bay Road via Kensington; then as above; line declared unsafe November 1887.
DUNEDIN CITY & SUBURBAN TRAMWAY COMPANY
Operated as an adjunct to its street tramway system.
DUNEDIN CITY CORPORATION
Only spasmodic services operated by New Zealand Railways with its own rolling stock, e.g, on show days. Line also used for horse sales at Tahuna Park and movements of mounted troops during the First World War.

The Dunedin, Peninsula & Ocean Beach Railway Company Limited

Steam Trams

Merryweather (Two tram locomotives) Merryweather No. 2 standard type.
Fleet Names: *Anglia* and *Scotia*.
Year Built: 1878
Purchased second-hand from the Wellington Tramway Company in December 1880. Former Wellington fleet numbers No 8, *Anglia*, and No. 7, *Scotia*.
Withdrawn: 31 May 1882
Builder: Messrs Merryweather and Sons, Greenwich, England.
Maker's numbers: 86 and 87
Specifications as built: Wheel arrangement: 0-4-0. 7 inches x 11 inches inside Cylinders. Wheel diameter: 2 feet. Wheel base: 4 feet 6 inches. Water condenser on roof. Tank capacity: 300 gallons. 10 h.p. boiler, rating 120 lbs psi.

The Dunedin City Council took over the assets of the DP&OBRC and its successors on 16 November 1901.

No. 86 used as a shunting locomotive at the Dunedin Gasworks 1901-1905. Then sold and used as a bush locomotive by the NZ Pine Company, Seaward Bush, Southland. Scrapped 1926.

No. 87 used as a shunting locomotive, Woolston, Christchurch 1902-1906, then sold to the Otatara Land Company, Invercargill, used as a steam-tram hauling passenger trailers – April 1907-1911. Then used as a construction locomotive on the New River Estuary reclamation, now known as the Oreti, and on the PWD Orawia Railway extension of the Tuatapere branch. Scrapped 1925.

The Merryweather tram locomotive *Anglia* with trailers in Crawford Street. Liverpool Street is in the background. The *Anglia* had been bought second-hand by the Peninsula and Ocean Beach Railway Company from the Wellington Tramway Company in December 1880. Four trailers had also been purchased second-hand from the Wellington company. These trailers built by John Stephenson Company of New York in 1878, seated 18 passengers. *Before the cars were run in Dunedin, the tramway wheels were exchanged for railway wheels in the interests of safe working.

*Jim Dangerfield research

This book only covers the rolling stock used on these lines when operated as a steam tramway with city street running for a brief period from January 1881 to May 1882. The book, *Dunedin's Matchbox Railway,* by J.A. Dangerfield, published in 1986 by the New Zealand Railway and Locomotive Society, Wellington, tells the full story of this unique railway-tramway venture.

DUNEDIN
Maori Hill
Horse Tramway

The Roslyn Tramway Company Limited

From shortly after the Roslyn cable line opened in February 1881 to 23 October 1900

Gauge 3 feet 6 inches
The population of Roslyn was 2875 and Maori Hill 1131 was when the system opened.

The Suburbs and Streets Served by the Horse-Tram

From the Roslyn cable tramway terminus (also known as the Junction), in Ross Street (Town Belt), via Scarba Street, City Road to Roslyn Council Chambers (High Street); line subsequently extended along High Street (Highgate) to Bishopscourt. Horse tramway car barns and stables erected on the site of the present Roslyn Fire Station.

Above: This very early photograph of the original powerhouse of the Dunedin and Roslyn Tramway Company at the Town Belt shows a two-horse Maori Hill horse-tram on the right and one of the original Rattray Street cable cars by the powerhouse.

Left: On the opening day of the Maori Hill electric tram service in October 1900, this photograph shows a City and Suburban Tramway Company car No. 108, with the city company's name still on the letter board panel at the Town Belt terminus of the Roslyn company lines. *Russell Grigg collection*

DUNEDIN
Roslyn
Cable Tramway

The Roslyn Tramway Company Limited

(later named The Dunedin & Roslyn Tramway Company Limited)
From 6 February 1881 to February 1921

Dunedin City Corporation Tramways

From 31 January 1921 to 14 August 1939

Dunedin City Corporation Transport

From 14 August 1939 to 26 October 1951

Gauge 3 feet 6 inches
The population of Roslyn was 2875 when the system opened

The Suburbs and Streets Served by the Cable Trams

Original line: From the junction of Rattray and Maclaggan Streets up Rattray Street and through the Town Belt (reserve track) to terminus near Ross Street. Single track with two passing loops near York Place and Smith Street (by St. Joseph's Cathedral).
Route length: Approximately 77 chains. **Maximum gradient:** 1 in 6.5. **Rope length:** 7000 feet.
Rope speed: 6 mph. Line duplicated 1884.

Early c1900 Extensions:
(1): From Town Belt terminus via Ross Street, Highgate, Belgrave Crescent (double track), and James (now Falcon) Street to Castle Street (now Frasers Road), Kaikorai Valley terminus (single track). The extension via Belgrave Crescent and James Street was not licensed for passenger traffic by the PWD due to concern over safety standards. However, until 1906 the conduit in James Street was used to convey the traction rope for operating the Princes Street to Belgrave Crescent section of the line, a temporary terminus was established at Highgate. The James Street section was abandoned in 1906.
(2): Extension down Rattray Street from the Maclaggan Street intersection to Princes Street.
Route length: 1 mile 40 chains. **Maximum gradient:** 1 in 3.5. **Vertical lift:** 600 feet. **Rope length:** Approximately 15,000 feet. **Rope speed:** 6 mph.

August 1906 Extension:
Double-track line from Highgate into Kaikorai Valley via a deep cutting and reserved right-of-way to Frasers Road terminus powerhouse and car barn. Grip cars fitted with wheel-operated Fell rail emergency braking system for this section.
Route length: 1 mile 30 chains. **Maximum gradient:** 1 in 4.7. **Vertical lift:** 660 feet. **Rope length:** 15,700 feet. **Rope Speed:** 6 mph. In 1911 increased to 7.95 mph.

The Roslyn (Rattray Street) line was the last steam-driven endless wire rope cable tramway in the world.

Dunedin & Roslyn Tramway Company

Cable-Trams

Fleet Numbers: 1 to 4 – first series to use these road numbers. (four grip trams)

Year built: 1880

Withdrawn: Early 1900s – one retained stripped of seats for the cartage of coal.

Builder: Unknown – "dummy cars" (grip cars) were recorded in the *Otago Daily Times*, June 1880, as having been "...ordered from America and the rails from Home..."; later reported: "cars arrived early in the New Year of 1881 on board the ship *Mabel* ".

Specifications as built: Four-wheel truck. Length over body about 14 feet. Passengers seated: 16 + standees totalled 25. Gripper: *Duncan wheel-operated screw type, double jaw (bottom) side grip. Brakes: lever-operated wheel; lever-operated slipper although at least one car fitted initially with screw-operated track brake. Screw-operated emergency "dolphin" brake fitted post-May 1881. Lever-operated dolphin brake is still used for emergencies on the San Francisco cable cars.

Open cable trams with longitudinal and cross seating on four sides of the centre located gripman's compartment. Originally uphill end (facing Roslyn) was glass, later downhill (facing City) was glassed in for added passenger protection. Lighting by kerosene lamps. Trams fitted with hooks and chains for possible trailer haulage; this did not eventuate.

* According to Gerald Ditchfield research.

No. 4, one of the original American-built Roslyn cable trams at the city terminus at the corner of Rattray and Maclaggan Streets, shortly after the route opened in 1881. The gradient ahead is 1 in 7.5; note hooks and chains for potential trailer haulage. *Evening Star*

Management and traffic staff of the first cable tramway in the world to be built outside the USA. The cable grip tram is at the Town Belt terminus. The car is sitting on the "transfer table" which provided manual lateral access from the car shed (side entrance) onto the tram line. *Hocken Library*

By the time this photograph was taken, both ends had been rebuilt with glassed-in fronts and advertising boards placed on the roof.

Above: One of the original four grip trams after new glass fronts had been fitted and advertising boards placed on the roof. The cable grip tram was standing at the original city terminus at the corner of Rattray and Maclaggan Streets. The original cable tension gear was located in a vault below the surface of the roadway in front of the car in the photograph. *Jim Dangerfield*

Below: The same view over 60 years later – one month before the service finished. A city-bound cable-tram rounding the corner in Rattray Street by the intersection with Maclaggan Street . The city terminus was extended down Rattray Street to Princes Street in early c1900. *Graham Stewart*

Above: A cable-tram standing at the original Rattray Street city terminus by the corner of Maclaggan Street when only the uphill ends of the original grip trams were glassed in. The gripman is leaning against the original wheel-operated screw-type gripper. The Crown Hotel features on the right in both photographs on this page. *David Todd*
Below: Looking down Rattray Street to Princes Street from the same corner in September 1951 – the cable tracks from this corner had been originally extended down to Princes Street as a single track on the right-hand side of the road in early c1900, then at a later date the track was duplicated. *Graham Stewart*

Cable Coal Tram

Fleet Number: Not known (one coal grip tram)
Year built: 1881
Withdrawn: Unknown
Builder: Unknown
Used to haul bagged coal from the city to the steam-driven Town Belt powerhouse and later to the new steam-driven powerhouse in the Kaikorai Valley.
Specifications as built: Four-wheel truck, length approximately 14 feet, open, no passenger seating. Grip and braking mechanism believed to be the same as No's 1 to 4 grip trams.

No plans have been found.

This photograph, published in the *Auckland Weekly News* in October 1927, shows the Roslyn cable line grip coal tram with five of the final fleet of cable grip trams built between 1903 and 1908 for the company. The Roslyn Tramway Company's car shed and powerhouse in the Kaikorai Valley is in the background. The original Duncan wheel-operated screw-type grip shows clearly on the "coal" tram in the foreground. *Denton*

One of the original 1881 cable grip trams by the curve outside St Joseph's Cathedral in Rattray Street. This curve was a first for cable traction in the world. Dunedin-born 29-year-old engineer George Duncan designed the "pull curve", a series of small wheels, known as "drum pulleys", which gently eased the wire rope around the curve and still allowed the tram to grip while taking the 1-7.5 upward grade. It was from this left-hand curve that the spectacular fatal runaway occurred in April 1881. *Russell Grigg collection*

Two of the 1881 American-built cable grip trams at the Town Belt terminus in October 1900. A combination cable grip tram is parked on an elevated inspection track located at the city end of the terminal complex in the background. *Russell Grigg collection*

Cable Water Tram

Fleet Number: Not known (one grip tram)
Year Built: 1881
Withdrawn: When the water supply to the powerhouse was established. Parts may have been used later for the construction of passenger grip tram No. 5.
Builder: Not known
Specifications as built: Four-wheel truck, open tram, no passenger seating, fitted with large tanks to convey water from the city terminus to the Town Belt steam powerhouse.
Gripper and braking believed to be the same as No's 1 to 4 grip trams.

No plans or photographic records found.

Cable Experimental Tram

Fleet Number: Not allocated.
Year Built: c1881-1882
Designer: G.S. Duncan.
Builder: George & McKenzie.
Specifications as built: Constructed as a mobile iron frame only, believed fitted with two screw-type (Duncan?) grips at each end of the frame to test the feasibility of double-ended cable-tram operation. Found impractical due to severity of vertical curve located on the reserve track between Queens Drive overbridge and Arthur Street.

No plans or photographic records found.

Cable-Trams

Fleet Numbers: Never commissioned or given fleet numbers (four cable tram bodies only)
Year built: 1901
Builder: Bodies built in Melbourne.
Specifications: Designed by J. Meiklejohn.
Completely enclosed – never fitted with running gear.

The Public Works Department would not issue a licence to run as the bodies were considered to be too long for the gradient curves on the line.
No plans or photographic records ever found.

Cable-Trams

Fleet Numbers: 5 and 6 (two grip trams)
Year built: late c1880s
Withdrawn: No. 5 – c1906
No. 6 – retained, body stripped and used as a gripman's training tram and for the cartage of coal for a few years; no Fell brake fitted.
Builder: Not known
Specifications as built: Combination cable tram, closed saloon at each end with sliding doors with two cross-bench seats facing each other. In the centre on each side, knife-board back-to-back longitudinal seating, each side of the gripman's compartment or "well". Movable glass weather shields on either side of the "well" to protect the gripman. Gripper: Duncan screw type, operated double-jaw (bottom) side grip. Lacked ground clearance to be fitted with Fell brake to operate on the final extension into the Kaikorai Valley.

In 1902 No. 6 was fitted experimentally with a lever-type gripper based on a Melbourne design.

Test Trolley

Fleet Number: Not allocated
Year built: 1901
Withdrawn: Not known
Builder: Built privately by a Mr W.W. Richardson of Outram to test the feasibility of automatic wheel brakes for cable trams.
Specifications as built: A four-wheel truck test trolley which was manually propelled.

Trial runs were conducted on the James Street cable car tracks on 4 June 1901; on the reserve track, unformed section of the street. The *Otago Daily Times* were present to witness the trials and reported as being "impressed". No other record of this experiment has been found.

No. 5, the first combination cable grip tram built for the Roslyn line, had a closed saloon at each end with sliding doors for passenger protection. This photograph shows No. 5 outside the original Roslyn powerhouse at the Town Belt terminus. Behind the cable-tram can be seen the traverser which was used at this terminus to manually switch grip trams from the up-line (left) to the down-line (right). This form of working is believed to have continued until c1903. *Otago Settlers Museum*

Cable-Tram

Fleet Number: 1 – (one grip tram)
Second cable tram to use this fleet number.
Renumbered: No. 94 by the DCCT in the 1920s.
Year built: September 1903
Withdrawn: 26 October 1951
Builder: Cossens & Black, Dunedin.
Specifications as built: Designed by J.Meiklejohn. Four-wheel, wheel base: 5 feet 3¾ inches, enclosed saloons with sliding doors with two cross-bench seats facing each other at each end. In the centre on each side, knife-board centre back-to-back longitudinal seating with the gripman's compartment "well" between. No. 1 was a slightly larger tram than numbers 2 to 7 – only cable tram of this series to have a clerestory roof.
Height over cab: 9 feet 4½ inches, width over steps: 7 feet 2 inches, width over body: 6 feet 7¼ inches, length over bumpers: 18 feet 9 inches, length over body: 17 feet 3 inches, weight: 4.5 tons. Gripper: lever-operated double-jaw (bottom) side grip. Brakes: lever, independently operated wheel and track, wheel-operated Fell, ratchet & pawl "run-back" brake. Licensed to carry 35 passengers.
No. 94 was the second-to-last cable tram to leave the Exchange when the line closed on 26 October 1951.

No. 1 in 1906 with the distinctive clerestory roof. *Tramway Historical Society*

Although this "comfortable load" in 1925 has obscured the fleet number, it is No. 1, by now renumbered 94, running under the management of the Dunedin City Corporation Tramways. This was the only Roslyn grip tram built with a clerestory roof. The luggage carriers were later placed on the front of the grip cars to stop schoolboys jumping up and down at the back to deliberately cause the car to derail.

Above: No. 94 at "The Junction" stop after buses had replaced the Maori Hill electric tram in 1936. *Evening Star*

Right: The gripman "hooking" the rope into the gripper after No. 94 had used the cross-over for short working at Highgate. *Graham Stewart*

Cable-Trams

Fleet Numbers: 2 to 7 – (six grip trams)
Second series of cable-trams to use these fleet numbers.
Renumbered: No's 91 to 93, 95 to 97 by DCCT in the 1920s.
Year built: 1906 to 1908
PWD tests: No. 2 – 13 August 1906
No. 3 – 14 August 1906
No. 4 – 24 December 1906
No. 5 – 24 December 1906
No. 6 – 7 December 1908
Withdrawn: 26 October 1951
Builder: Cossens & Black, Dunedin.
Specifications as built: Four-wheel, wheel base: 5 feet 3¾ inches, enclosed saloons with sliding doors with two cross-bench seats facing each other at each end. In the centre on each side, knife-board centre back-to-back longitudinal seating with the gripman's compartment between.
Height over cab: 9 feet 4½ inches, width over steps: 7 feet 2 inches, width over body: 6 feet 7 ¼ inches, length over bumpers: 18 feet 9 inches, length over body: 17 feet 3 inches, weight: 4.5 tons. Gripper: lever-operated double-jaw (bottom) side grip. Brakes: lever, independently operated wheel and track, wheel-operated Fell brake, ratchet & pawl "run-back" brake.
All these six cables trams had arch roofs – some of the cars had slightly different front-window framing. Cable-trams 4 and 5 were similar in every respect to No's 1, 2, and 3 – with the exception of the saloon doors which were slightly narrower and instead of double-sliding, they were single-sliding doors.
*It is believed that during construction of this series of cars, a car frame was taken down to Joseph Sparrow's foundry in the city to be experimentally fitted with gripper and braking levers at either end to again test the practicality of double-ended operation. The gripping mechanism although located in the centre of the frame was connected to the operating levers at each end by rods and cranks. However, the frame never left the foundry in this form and was subsequently returned to the car barn in Kaikorai Valley.
*Gerald Ditchfield research

Above: No. 4 in company days outside the car shed and powerhouse in Castle Street (Frasers Road) in Kaikorai Valley. *Russell Grigg collection*

No. 92 was the last cable-tram to run from the city to Kaikorai Valley when the line closed on 26 October 1951. All cars of this class were in service on the last day except No. 93 which had a broken grip.

Above: Cable grip trams No's 2, 3, 5, and No. 1 at "The Junction" when a feeder electric tram service, also operated by the Roslyn Company, ran in conjunction with the Roslyn cable tram service to Maori Hill between 1900 and 1936. *Gerald Ditchfield collection*

Right: Cable grip tram No. 6 with the private company name still on the front apron, when a strap and a toe-hold was all that were required to be carried safely home to Roslyn.

KAIKORAI

KAIKORAI
25
P.153

THE ROSLYN CABLE LINE

Opposite Page
Above left: A Kaikorai Valley-bound cable grip tram ascending through the Town Belt after passing under the old wooden bridge in Queen's Drive. *Graham Stewart*

Left: Lunchtime passengers transferring from a motor bus to cable-car at "The Junction" while Otago Girls' High School pupils rush to board the car. *Graham Stewart*
This Page
Above: The dramatic change in gradient from near level to a 1in 4.7 downward grade into the Kaikorai Valley. The gripman would call out "Watch out for the dip". A Fell centre-rail braking system was used on this section which can be seen in the foreground. *Graham Stewart*

Right: A city-bound Roslyn cable-car climbing out of the Kaikorai Valley towards the cutting. The Fell brake rail can be seen on both tracks. *Graham Stewart*

Right: The last Roslyn cable grip tram, No. 92, arriving at the Kaikorai Valley terminus on the evening of 26 October 1951. *Stan Rockliffe*

In July 1969 the Tramway Historical Society obtained the body of Roslyn cable grip tram No. 95 from Waihola for restoration.

Right: Using the slight downgrade, No. 95 coasts away from the Princes Street city terminus with a comfortable load during the last month of service. The moving traction rope was "picked up" by the gripman some distance further on. *Graham Stewart*

Below: Being loaded on to a truck for the trip back to Christchurch. The society faced a major restoration task. *Cecil Bachop*

Below right: Fully restored to c1940, No. 95 is now an exhibit in the Hall of Wheels at Ferrymead, Christchurch. *Trevor Craib*

DUNEDIN
Mornington
Cable Tramway

The Mornington Tramway Company Limited
From 23 March 1883 to 1 April 1903

Mornington Municipal Tramways
From 1 April 1903 to 1 January 1916

Dunedin City Corporation Tramways
From 1 January 1916 to 14 August 1939

Dunedin City Corporation Transport Department
From 14 August 1939 to 2 March 1957

Gauge 3 feet 6 inches
The population of Mornington was 2886 when the system opened.

The Suburbs and Streets Served by the Cable Trams

"Main Line": From the intersection of Princes Street and High Street (Exchange), double track up High Street and Eglinton Road to Parkhill Avenue – reserve track in Eglinton Road – right-hand street operation.

Route length: 1 mile. **Maximum gradient:** 1 in 6.25. **Vertical lift:** 450 feet.
Rope length: "Spliced and running" 11,500 feet. **Rope speed:** 7.95 mph.
20 September 1925: Electric haulage installed and rope speed increased to 10.66 mph.
11 – 16 August 1928: Main line converted to left-hand street operation; temporary rope from Roslyn line used 16 August – 30 September 1928.

Mornington Tramway Company Cable-Trams

Fleet Numbers: Not known (four grip trams)
Year built: 1882 – entered service 21 March 1883.
Withdrawn: Destroyed by fire 9 February 1903.
Builder: Cossens & Black, Dunedin.
Specifications as built: Four-wheel, open cable trams with clerestory roofs, and an oil-lamp ventilator dome in the centre. Movable glass fronts fitted after service commenced together with a canvas cover which could be fitted over the wire-mesh front apron. Two cross-bench seats at each end of the tram faced out and on each side in the centre, a knife-board back-to-back longitudinal seat with the gripman's compartment "well" in the centre. Brakes: lever, independently operated wheel and track, also ratchet & pawl "run-back" brake. Licensed to carry 21 passengers. Gripper: lever operated, *Henry Root (USA) type single-jaw (bottom) side grip; modified by the Mornington Tramway Company engineer, J.Louden, in 1888.

In the 1900s these grip trams were rebuilt with both ends being lengthened to accommodate more seated passengers and the arched roofs extended the full length of the tram, square-ended glass fronts fitted. This resulted in larger car loadings and with a worn cable, the powerhouse engine was sometimes overloaded on wet nights, forcing passengers to walk home.

* From Gerald Ditchfield research.

Above: An original Mornington grip tram with trailer No. 1 at the Mornington terminus. The wooden treads set into the road on each side of the track in the foreground were to give staff traction when manually moving the trams. Note the large gripper quadrant. *J.Walters, Trevor Craib*
Right: Two of the original grip trams after removable glass fronts were fitted to one end shortly after the service commenced. The cars followed American practice of right-hand street operation.

Left: Another view of these tiny Dunedin-built cable grip trams. They would have been the first passenger cable-trams to be built in New Zealand. The clerestory roof would have been for decoration only on a completely open tram.

Left: By the time this photograph was taken at the Mornington terminus, both ends of the cable grip tram had been fitted with removable glass fronts. The local constable stands in front of the trailer fleet number, always a frustration for tramway historians. *Otago Settlers Museum*

Below: Cable grip tram No. 7 with trailers after the grip trams had been lengthened c1900, at both ends to accommodate more seated passengers. *Otago Settlers Museum*

Trailer Cars

Fleet Numbers: 1 to 4 (Four "bobtail" type trailers)
Year built: 1882 – entered service May 1883
Withdrawn: Three destroyed by depot fire 9 February 1903.
The remaining trailer, No. 4, continued in service –
Renumbered: No. 107 by the DCCT in the 1920s.
Builder: Jones Car Company, New York, USA.
Specifications as built: Standard horse-tram design. Four-wheel truck, single saloon with clerestory roof, longitudinal seating, saloon sides; curved lower rocker panel, straight upper waist panels, drop sash windows, end platforms with open fronts. Brakes: Hand wheel and hand track - operated from either end, also fitted with ratchet & pawl "run-back" brakes; could be activated by gripman. No life guards. Length: 16 feet 4 inches, width: 6 feet 9 inches, height: 9 feet, weight: 33 cwt.

*When the cars arrived in Dunedin, one trailer is believed to have been lent to the Dunedin & Roslyn Tramway Company to test the suitability of hauling trailers on their original Rattray Street line; these tests were not successful.
The body of No. 107 was purchased from a private owner by the Old Time Transport Preservation League of Matakohe, North Auckland, and shipped north to Auckland in 1958 for preservation. Now an exhibit at the Museum of Transport and Technology, Western Springs, Auckland.

*Gerald Ditchfield research

Above left: No. 2 trailer as built by the Jones Car Company of New York. *Above:* A few years later a solid panel had been fitted to replace the netting front apron and roof advertising boards fitted. A Maryhill composite grip car was providing haulage on the High Street line. *Ray Hargreaves*

The four American-built trailers photographed at the Mornington terminus with two of the original grip cars in September 1900. More roof advertising boards had been added to some of the trailers. *Otago Witness, Cecil Bachop*

No. 4 trailer has certainly been a survivor – it was the only piece of rolling stock to survive the fire that destroyed the Mornington Tramway Company's tramway sheds and plant on 9 February 1903. Then fifty years later, after being dismantled and sold to a private buyer for use as a shed, was rescued for the second time for preservation by Mervyn Sterling of the Old Time Transport Preservation League of Matakohe, Northland, on a visit to Dunedin.

Left: The morning after the 1903 fire with the paintwork badly scorched but intact. No. 4 (107) was to give another 50 years of service to the suburb of Mornington. *Otago Witness, Ray Hargreaves*

Above left: Renumbered No. 107 as part of the Dunedin City Corporation's fleet, on the trailer siding in Parkhill Avenue, Mornington, in the early 1950s. *Gerald Ditchfield*

Above: Being unloaded from the SS *Waitemata* on arrival at Auckland in September 1958. *Graham Stewart*

Left: Now partly restored as an exhibit at the Museum of Transport and Technology, Auckland. *Bruce Pullar*

Trailer Coal Car

Fleet Number: 8 (one coal trailer)
Renumbered: No. 108 by the DCCT in the 1920s.
Year built: 1891
Withdrawn: After steam engine haulage was replaced by electric drive in 1925.
Builder: Mornington Tramway Company.
Specifications as built: Similar in size to the trailers. Only fitted with a hand-operated wheel brake. Bagged coal carried: 1.5 tons. On occasions was fitted with portable passenger seats.

*Used by the company to make up three-car "trains" (one grip tram and two trailers) during the 1889-90 Dunedin Exhibition period. PWD later refused to license this type of operation. Coal trailer also used in 1907 as a single passenger trailer during morning peak periods.

**Trailer involved in a spectacular runaway in 1911 whilst unloading coal into the powerhouse bins from the Maryhill Extension line. Car accidentally broke away and caught up with the regular service out-bound grip car part way down the 1 in 3.5 gradient. Both the trailer and the grip tram badly damaged; two employees were injured. The coal trailer was subsquently anchored to the car barn wall when carrying out this operation.

Fleet number taken over by a new cable tram built in 1925.

In 1891 according to the company minute book*** – the company built a passenger and goods cable-tram with seating for 26. No other records exist to confirm or deny this minute book entry.

* Gerald Ditchfield research.
** Gerald Ditchfield research.
****Straphangers and Grippers,* Bill Campbell & Ray Hargreaves, Otago Heritage Books 1994.

The only known photograph of the trailer coal car seen at the city terminus in High Street. Sadly, the only identifiable sign-writing on the trailer are the words "Municipal Tramways" An enlarged section of a picture postcard of the period which fortunately for tramway historians included this unique tram. *Russell Grigg collection*

Mornington Municipal Tramways

Cable-Trams

Fleet Numbers: 1 to 4 (four grip trams)
Fleet Numbers originally used by the 1882 trailers.
Renumbered: No's 101 to 104 by the DCCT in the 1920s. DCCT added 100 to each of the fleet numbers.
Year built: entered service No's 1 – 6 July 1903.
2 – July 1903.
3 & 4 – December 1903.
Withdrawn: 2 March 1957
Builders: Glaister & Carey, Dunedin. Ironwork: Joseph Sparrow & Sons, Dunedin.
Designer: Mr Roberts
Specifications as built: Four-wheel, cable grip trams, enclosed saloons with sliding doors with two cross-bench seats facing each other at each end. In the centre on each side, knife-board centre back-to-back longitudinal seating with the gripman's compartment "well" between, clerestory roof.
Height: 8 feet 9 inches, width: 7 feet 10 inches, length: 19 feet, weight: between 5 & 6 tons. Brakes: lever, independently operated wheel and track, also ratchet & pawl "run-back" brake. Licensed to carry 38 passengers.

In 1911 the levers that operated the Lowden grip were reversed, i.e, pushed forward to take hold of the cable during downhill travel, pulled back during uphill travel. Size of gripper quadrant also reduced.
After the depot fire destroyed all the rolling-stock except for trailer No. 4 (later No. 107), two uncompleted new cable-trams were fitted with temporary seating and a canvas framed roof. The first being placed in service on 25 April 1903, and the second three days later, until new cable-trams were available to be commissioned. These two cable-trams (to be given the fleet numbers 3 & 4) were then withdrawn and taken back to Glaister and Carey for the coachwork to be completed.
Years later the original five-light wooden-framed front windows were replaced with three-light steel-framed windows.

Left and above: The two temporary cable-cars with canvas roofs which were placed in service to maintain a basic 15-minute service after the fire in April 1903. *Jim Dangerfield / David Todd*

Left: No. 1 cable-grip tram, the first to be placed in service in completed form after the 1903 fire. *Cecil Bachop collection*

Left: No. 1 seen a few years later, repainted with more decorative paint work and extra footboard. *Tramway Historical Society*

Above: No. 3 at the city (Exchange) terminus in High Street. The design of the locally built Dunedin cable-trams enabled the ladies in their long dresses to board the tram with ease, 90 years ahead of the low-floor trams which were introduced to European electric tramway systems in the 1990s.

Above and right: When it was fashionable to travel by cable-tram. Two Muir & Moodie postcards of High Street in the 1900s. The car (above) in the foreground is No. 1 and at right, No. 4. *Ray Hargreaves / Trev Terry*

Below: No. 1 and the rest of this series had the front windows replaced in later years with three-light steel-framed windows. *Graham Stewart*

Dunedin City Corporation Tramways

Cable tram

Fleet Number: 108 (one grip tram)
Year Built: 1925
Withdrawn: 14 April 1956
Builder: Dunedin City Corporation Tramways.
Specifications as built: Four-wheel, cable grip tram, enclosed saloons with sliding doors with two cross-bench seats facing each other at either end. In the centre on each side, knife-board centre back-to-back longitudinal seating with the gripman's compartment between, arched roof. Designed by J. Meiklejohn.
Height: 8 feet 7 inches, width: 7 feet 3.5 inches, length: 18 feet 3 inches, weight: about 5 tons. Brakes: lever, independently operated wheel and track, also ratchet & pawl "run-back" brake. Licensed to carry 38 passengers.

No. 108 was the last cable-tram to be built for service on the Mornington line.
Withdrawn from service after being damaged in collision with a truck on 14 April 1956.

Above: No. 108 and trailer No. 112 climbing the High Street gradient with ease in September 1951. *Graham Stewart*

Left: A collision with a motor truck on 14 April 1956 was the final terminus for No. 108. As the service was scheduled to close within a year, management decided it would not be economic to repair the grip tram. *Ted Helm collection*

Trailer Cars

Fleet Numbers: 9 and 10 (two "bobtail" type trailers)
Renumbered: No's 109 and 110 by the DCCT in the 1920s.
Year built: 1880s
Withdrawn: 21 January 1957
Builder: Unknown – probably imported.
*The running gear had the following cast into the wheel spring cups "Veebe (?), New York, 3 Dec 1878".
Rebuilt by Mornington Municipal Tramways.
Specifications as built: Four-wheel truck, standard horse-tram design, five window single-saloon with ventilated arched roof, longitudinal seating, saloon sides; straight lower rocker panels, straight upper waist panels, drop sash windows, end platforms with open fronts. Brakes: Hand wheel and hand track – operated from either end; also fitted with ratchet and pawl "run-back" brakes; could be activated by gripman. No life guards.

The Evening Star, 10 November 1910 reported the following "The last of the DCC horse cars have been sold. Two have been purchased by the Mornington Borough Council and it is understood that it proposes to alter them and run them as trailers on its cable car line."

*From Gerald Ditchfield research.

Below: No. 109 trundling down High Street behind cable grip tram No. 101 for the late afternoon peak period traffic in 1951. *Graham Stewart*

Right: No. 110 all alone – waiting at the city terminus while cable grip tram No. 104 runs ahead through the cross-over, before taking hold of the up-bound cable to move ahead to allow the trailer car to run in behind onto the up-track, for yet another trip up the High Street grade to Mornington. *Graham Stewart*

Trailer Cars

Fleet Numbers: 111 and 112 (two "bobtail" type trailers)
Year built: No. 111 PWD tested 24 December 1920
No. 112 entered service 13 September 1924
Withdrawn: 21 January 1957
Builder: Dunedin City Corporation Tramways.
Specifications as built: Four-wheel truck, 24 inch diameter wheels, four window single-saloon with ventilated arched roof, longitudinal seating, saloon sides; curved lower rocker panels, curved upper waist panels, drop sash windows, end platforms with open fronts. Brakes: Hand wheel and hand track – operated from either end, also ratchet & pawl "run-back" brake which could be activated by the gripman. No life guards.
Length: 17 feet 3 inches, width: 7 feet 1 inch, height: 9 feet, weight: 35 cwt. Seated 14.

Above: No. 112 running solo on to the up-line at the city terminus in High Street. *Graham Stewart*

Above: No. 111 being hauled by cable-grip tram No. 103, passing Grant Street on the downward grade in High Street in 1951. *Graham Stewart*

Left: No. 112 on the same afternoon, being hauled by cable grip tram No. 108, bound for the city to collect the waiting rush hour crowds for their homeward journeys where an evening meal would be waiting. *Graham Stewart*

The first duty of the day for Mornington cable-tram crews was to push the cars from the car sheds. This series of photographs was taken in January 1955.

Above: Trailer No. 112 is the first out of the sheds, being pushed across Elginton Road to the trailer siding in Parkhill Avenue. *Gerald Ditchfield*

Above right: The next task was to roll a grip car (No. 108) out of the car shed and onto the turntable. *Gerald Ditchfield*

Right: A bit of "bum shunting" could be involved when turning the grip car on a 90° angle to line up with the main line downhill track. *Gerald Ditchfield*

The trailer-holding siding in Parkhill Street opposite the Mornington cable tramway car sheds in December 1955. From left, the trailers are No107, now preserved at the Museum of Transport and Technology, Auckland, No. 110, No. 112 and No. 109. A cable-grip tram can be seen in the doorway of the car shed. *Gerald Ditchfield*

To reposition a grip tram and trailer at the city terminus of the Mornington line in High Street after a downhill trip, the trailer had to be shunted.

Top: The grip tram had stopped by Manse Street to detach the trailer, and is shown in the photograph having released and thrown the rope clear of the grip, to coast down High Street and is about to cross over onto the uphill track. *Gerald Ditchfield*

Above: The grip tram has now moved ahead on the uphill track to clear the cross-over while the conductor, operating the goose-necked hand brakes, coasts the trailer down and over the cross-over onto the uphill line. *Gerald Ditchfield*

Left: The trailer has now been safely parked and the grip tram is running back – the grip being held in an open position – the conductor is walking forward to couple the trailer with specially designed link and pin-type coupling and heavy safety hooks and chains to the grip tram. *Gerald Ditchfield*

DUNEDIN
Maryhill Extension
Cable Tramway

The Mornington Tramway Company Limited

From 18 March 1885 to 1 April 1903

Mornington Municipal Tramways

From 1 April 1903 to 1 January 1916

Dunedin City Corporation Tramways

From 1 January 1916 to 14 August 1939

(Service suspended between 1916 and 1921)

Dunedin City Corporation Transport Department

From 14 August 1939 to 29 October 1955 (for passenger traffic)
Officially closed 24 February 1956

Gauge 3 feet 6 inches
The population of Mornington was 3334 when the system opened.

The Suburbs and Streets Served by the Cable Tram

From the Mornington Township, adjacent to Eglinton Road and Parkhill Avenue, (outside the right-hand side of the cable tramway powerhouse and car shed), down a reserved track right-of-way then along Glen Avenue (now Glenpark Avenue), to Mitchell Avenue, Maryhill. Single track with a passing loop between Springhill and Neidpath Roads.

Route length: 40 chains. **Maximum gradient:** 1 in 3.5.
Rope length: "Spliced and running" 6,000 feet. **Rope speed:** 7.95 mph.

MARYHILL EXTENSION
Mornington Tramway Company
Cable trams

Fleet Numbers: 9 and 10 (two grip trams)
Year built: 1882
Withdrawn: Destroyed by the depot fire 9 February 1903
Builder: Unknown
Specifications as built: Four-wheel, cable grip trams, enclosed saloons at each end, single sliding door on each side, with a cross-bench seat facing each end. In the centre on each side, knife-board back-to-back longitudinal seating with the gripman's compartment "well" between in the centre, arched roof with a clerestory which ran the full length of the tram, with an oil-lamp ventilator dome in the centre, square-ended glass fronts.

Built as "composite" cable-trams able to operate on both the High Street and Maryhill lines – a single-jaw (bottom) side grip was used for the High Street "mainline" and a double-jaw (bottom) side grip for the Maryhill Extension – the change of grip was a manual task requiring the cable tram to be taken out of service.

Above: Near Henderson Street, No. 10, one of the original grip trams built for the Maryhill Extension in transit on this line's 1 in 3.5 gradient. The photograph was taken from Mornington, looking south down Glen Avenue towards the Maryhill Extension outer terminus. Cable-grip trams No's 9 and 10 were the first cable trams used on the Maryhill Extension line until their sudden destruction by fire in February 1903. *Hocken Library, Ray Hargreaves*

Left: No. 10 on the Maryhill Extension line at the top of the gradient by the powerhouse and car sheds at Mornington. It was from this point that Coal trailer No. 8 broke away in 1911. *Alexander Turnbull Library*

Composite Cable Tram

Fleet Number: 5 (one grip tram)
Renumbered: No. 105 by the DCCT in the 1920s.
Year built: c1904
Withdrawn: 2 March 1957.
Builder: Glaister & Carey, Dunedin.
Specifications as built: Four-wheel, cable-tram, enclosed saloons at each end, sliding door on each side, with a cross-bench seat facing each end. In the centre on each side, knife-board back-to-back longitudinal seating with the gripman's compartment between in the centre, clerestory roof.
Height: 9 feet, width: 7 feet 6 inches, length: 15 feet 6 inches, weight: about 6 tons. Brakes: lever, independently operated wheel and track, also ratchet & pawl "run-back" brake.

Composite cable tram No. 5 working the Maryhill Extension line during Mornington Municipal Tramway days. Note the wooden slot rail. *Russell Grigg collection*

No. 105 was a "composite" cable-tram which could be used on both the High Street line and the Maryhill Extension line. The gripper had to be changed depending on which line the grip tram was to operate. The gripper also had to be raised 3/8 of an inch for the High Street service to allow for the different heights of each route's ropes. The weight of this car made it unpopular with gripmen.
Placed in service on the Maryhill extension line early 1905 when it re-opened after the depot fire of 1903. Worked "the extension" until replaced by No. 106 in 1910. Then used between 1910 and 1925 as a spare cable-tram for both the High Street and Maryhill lines.
Last used on High Street 1956-57. When the High Street line closed in March 1957, this car was still painted in the Municipal colour of maroon with black and yellow lining and gold lettering. Presented to the Seashore Trolley Museum, Kennebunkport, Maine, USA, complete with grip. Shipped by the M.V. *Whangaroa* on 6 November 1957.

In the early 1950s No. 105 was working the Maryhill Extension as a relief cable-grip tram while the regular tram (No. 106) was undergoing maintenance. *Gerald Ditchfield*

No. 105's very last trip was on the day the High Street line closed on 2 March 1957. The man on the roof was there to strike the bell as the pull cord had broken. *Gerald Ditchfield*

Being unloaded at the Port of Maine, bound for preservation at the Seashore Trolley Museum, Kennebunkport, Maine, USA. *Foster M. Palmer*

DUNEDIN
Elgin Road Extension
Cable Tramway

Mornington Municipal Tramways
From 6 October 1906 to 22 January 1910

Service suspended until replacement ropes installed
9 October 1907 to 10 December 1907
25 July 1908 to 10 November 1908
(Replacement rope installed cSeptember 1909)
Service closed 22 January 1910
Some track points lifted 1911
Rope cut off at the powerhouse 1912, and left to lie in the street conduit

Gauge 3 feet 6 inches
The population of Mornington was 4150 when the system opened.

The Suburbs and Streets Served by the Cable Tram

From the Mornington Municipal Tramway Company's powerhouse and car barn in Eglinton Road, single track up Mailer Street, left into Lawrence Street (double-track passing loop), right into Havelock Street (single track), left into Elgin Road (double-track passing loop), line (single track) followed Elgin Road on right-hand side before crossing to left-hand side of Elgin Road between Bowler Avenue and Argyle Street before continuing to Balaclava terminus at Mitchell Avenue.

Route length: 78 chains. **Maximum gradient:** 1 in 8.
Rope length: Approximately 6550 feet. **Rope Speed:** 7.95 mph.

Believed to have been the last complete endless wire-rope cable tramway in the world to be constructed.

*It is believed that another passing loop may have also existed along Elgin Road between Bowler Avenue and Argyle Street, but this cannot be confirmed.

Elgin Road Extension Cable-Tram

Fleet Number: 6 (one grip tram)
Renumbered: No. 106 by the DCCT in the 1920s
Year built: 1906
Withdrawn: 29 October 1955
Builder: Coachwork by David Stansfield, Mornington. Steel frame built by the Mornington Borough Council.
Specifications as built: Four-wheel, mounted on axle boxes, cable tram, enclosed saloons at each end, sliding door on each side, with a cross-bench seat facing each end. In the centre on each side, knife-board back-to-back longitudinal seating with the gripman's compartment "well" between, arched roof. Curved end bulkheads. Height: 9 feet, width: 7 feet 6 inches, length: 16 feet, weight: about 4-5 tons. Gripper: Lowdon? double-jaw (bottom) side grip. Brakes: lever, independently operated wheel and slipper, fitted with bi-directional ratchet & pawl "run-back" brake for use on Maryhill extension. 22 passengers.

Built for the Elgin Road line, then transferred to the Maryhill line in 1910 when Elgin Road closed; *said to have briefly worked the Mornington "main line" in 1908. Except for temporary closures between 1907 and 1921 and maintenance relief by No. 105, worked the Maryhill extension almost continuously until line was made idle on 29 October 1955. Line closure made official on 24 February 1956.

No. 106 (minus gripper mechanism) is now an exhibit at the Otago Settlers Museum.

**Strap hangers & Grippers.* Bill Campbell & Ray Hargreaves

Right: No. 6 being prepared to be towed along the just completed Elgin Road extension cable line in 1906 for the testing of the track and curves. In the foreground work was proceeding on new trackwork requirements for the service outside the Mornington car sheds. *Otago Settlers Museum*

Below: In mint condition, No. 6 was photographed on the Elgin Road line for the *Otago Witness* on 10 October 1906. In the group are Messrs F.C. Ferry, the contractor (3rd from left); E.T. Clarke, chairman of the Council Tramway Committee (4th from left); H.A. Le Cren, mayor of Mornington (6th from left); and George Thom, acting manager of the Mornington Municipal Tramways (far right). *Ray Hargreaves collection*

Left: Renumbered 106, the front apron panel is sign-written "D.C.C.T.; No. 106, Maryhill". Working the Maryhill line in Dunedin City Corporation Tramways days. The Maryhill car always operated as a one-man tram, the gripman collecting fares usually whilst the tram was in motion. *Trevor Terry collection*

No. 106 outside the Glenvue Dairy at the Glenpark (Maryhill) terminus during a fall of snow in the 1930s. *Hocken Library*

No.106 descending the grade from the Mornington car sheds, bound for the Maryhill outer terminus in September 1951. *Graham Stewart*

No. 106 descending 1 in 3.5 grade in 1955. The curved arch front window panels had been replaced.

No. 106 is now an exhibit at the Otago Settlers Museum, Dunedin. *Graham Stewart*

DUNEDIN
Stuart Street
Cable Tramway

The Dunedin & Kaikorai Tram Company Limited

From 6 October 1900 to 23 May 1939
(Revenue service started 22 December 1900)

Dunedin City Corporation Transport Department

From 23 May 1939 to 31 July 1947

Gauge 3 feet 6 inches
The population of Roslyn was 4632 when the system opened.

The Suburbs and Streets Served by the Cable Trams

From the Octagon, double-track street running up Stuart Street and Albert Street, with reserve track through the Town Belt, with stops at Littlebourne, The Rectory, and Highgate. Descent into the Kaikorai Valley via Taieri Road – street running to Nairn Street terminus.
At Highgate the double-track cable line crossed the single-track electric Maori Hill tram line. Crossing later protected by railway-type warning lights and bells.

Route length: 1 mile 17 chains.
Maximum gradient: 1 in 4 originally, later eased to about 1 in 5.
Rope length: 12,605 feet. **Rope speed:** 7.95 mph.

Dunedin & Kaikorai Tram Company
Cable-Trams

Fleet Numbers: 1 to 5 (five grip trams)
Renumbered: No.'s 121 to 125 by the DCCT in the 1939
Year built: 1900
Withdrawn: 31 July 1947
Builder: Unknown
Specifications as built: Four-wheel, cable-trams, enclosed saloons with sliding doors with two cross-bench seats facing each other at each end. In the centre on each side, knife-board centre back-to-back longitudinal seating with the gripman's compartment "well" between, clerestory roof.
Height: 8 feet 9 inches, width: 7 feet 10 inches, length: 19 feet, weight: between 5 & 6 tons. Brakes: lever, independently operated wheel and track, also ratchet & pawl "run-back" brake. Gripper: Roberts (patent) double-jaw (top) side grip.

Because no provision had been made to accommodate grippers on car barn storage tracks, grips had to be removed and reinstalled in cars whenever they entered or departed the car barn.
The grips on these cars were difficult to work and often the conductor would have to "lend weight to the cause!"

Left: Cable grip trams No's 3 and 1 about to cross Queens Drive in the Town Belt when the service opened.

Below: No. 3 waiting for departure time at the Octagon. *Otago Settlers Museum*

Above: This photograph of No. 2 shows clearly the original fronts with the five glazed windows and the arched roof. Later the roofs were extended to give more weather and sun protection and the windows replaced with three large ones for better vision. *Gerald Ditchfield collection*

No. 4 arriving at the Stuart Street line terminus in the Octagon. *S.C. Smith, Tramway Historical Society*

Cable-Tram

Fleet Number: 6 (one grip tram)
Renumbered: No. 126 by the DCCT in 1939
Year built: c1938
Withdrawn: 31 July 1947
Builder: Dunedin & Kaikorai Tramway Company (Gus Brown). Steel frame built by McGregors.
Specifications as built: Four-wheel, cable grip tram, enclosed saloons with sliding doors with two cross-bench seats facing each other at each end. In the centre on each side, knife-board centre back-to-back longitudinal seating with the gripman's compartment "well" between. Only Kaikorai cable-tram with an arched roof – hence the nickname *Whaleback.*
Height: 8 feet 9 inches, width: 7 feet 10 inches, length: 19 feet, weight: between 5 & 6 tons. Brakes: lever-operated wheel and track also ratchet & pawl "run-back" brakes. Gripper: *R.L.Hay(?) double-jaw (top) side grip.

Because of the steel frame, gripmen regarded this car as "tricky" to handle.
No. 126 was the last cable-tram to be built for service in Dunedin.

* From Gerald Ditchfield research.

Left: No.6 in private company days at the Octagon terminus at the foot of Stuart Street, near the monument to Robert Burns.

Above: The body of No. 126 being transported on a motor truck when the cable-grip trams were sold after the line closed in July 1947. *Evening Star*

Left: Now part of the Dunedin City Corporation Transport fleet and renumbered 126 in the early 1940s. *Ross McQueen*

Left: The city terminus of the Stuart Street cable tramway was in the Octagon by the bronze staue of the Scottish poet Robert Burns. This photograph was taken from the steps of St Paul's Anglican Cathedral. The cable tram is No. 126. *A.P. Godber*

Above: No. 125 is seen leaving the reserve track section and crossing London Street on an inward run to the Octagon during the last month of service in July 1947. *Stan Rockliff*

Left: No 2 in private company days is about to cross the Highgate intersection where New Zealand's only cable / electric tramway crossing was located between 1900 and 1936. The crossing was protected by flashing lights and bells actuated by the up cable tram following a collision between a cable and electric tram in 1905.

Right: Cable-car No. 125 at the Littlebourne stop, about to cross Littlebourne Crescent before climbing the gradient to Highgate *Stan Rockliff*

Below: The "Whaleback" No. 126 ascending the 1 in 5 gradient to Highgate loaded with Otago Boys' High School pupils. *Evening Star, Gerald Ditchfield collection*

Lower: No. 126 and spare car, No. 3 parked behind, at the Kaikorai Valley terminus in Nairn Street in May 1947. *Stan Rockliff*

INVERCARGILL

Horse Tramway

Southland Tramway Company Limited

From 3 December 1881 to 1 October 1886

The Invercargill & Suburban Tramway Company Limited

From 1 October 1886 to 15 March 1908

Gauge 3 feet 6 inches

The population of Invercargill was 6774 when the system opened.

The Suburbs and Streets Served by the Horse-Trams

From the Post Office in Dee Street to Gladstone, along North Road December 1881
Along Tay Street and Conon Streets, to Ettrick Street December 1883
Branch line along Tay Street and East Road, to Mary Street 1883
Extension to Biggar Street, along Conon Street February 1884
Gladstone to Waikiwi October 1884

The service ran under different managements until a fire destroyed the stables, tram barn and four trams. The *Southland Times* of 16 March 1908 reported the last trip:

> Invercargill's last tram ran its last trip last night (Sunday), and such of the citizens of Invercargill who happened to be about at the time, made it a joyous farewell. As many as could went inside, a few sat on the roof, some more on the platforms, and the rest cheered. Possibly there were nearly 200 people in the post office square as the old car started grinding along in the direction of Waikiwi. There were no visible tears, but perhaps some of the merrymakers felt a shadow of regret at the sight of that shaky old servant stumbling away to the oblivion that waits for those who have passed their usefulness.

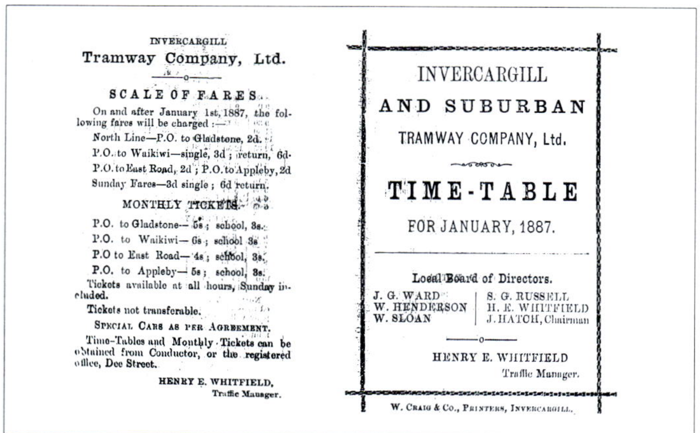

INVERCARGILL
Tramway Company, Ltd.

SCALE OF FARES.

On and after January 1st, 1887, the following fares will be charged :—

North Line—P.O. to Gladstone, 2d.
P.O. to Waikiwi—single, 3d ; return, 6d.
P.O. to East Road, 2d ; P.O. to Appleby, 2d
Sunday Fares—3d single ; 6d return.

MONTHLY TICKETS.

P.O. to Gladstone—5s ; school, 3s.
P.O. to Waikiwi—6s ; school 3s
P.O to East Road—4s ; school, 3s.
P.O. to Appleby—5s ; school, 3s.

Tickets available at all hours, Sunday included.

Tickets not transferable.

SPECIAL CARS AS PER AGREEMENT.

Time-Tables and Monthly Tickets can be obtained from Conductor, or the registered office, Dee Street.

HENRY E. WHITFIELD,
Traffic Manager.

INVERCARGILL
AND SUBURBAN
TRAMWAY COMPANY, Ltd.

TIME-TABLE
FOR JANUARY, 1887.

Local Board of Directors.

J. G. WARD | S. G. RUSSELL
W. HENDERSON | H. E. WHITFIELD
W. SLOAN | J. HATCH, Chairman

HENRY E. WHITFIELD
Traffic Manager.

W. CRAIG & Co., PRINTERS, INVERCARGILL.

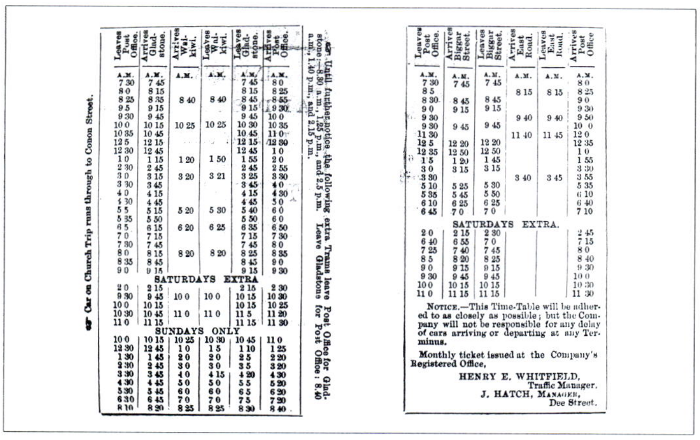

☞ Car on Church Trip runs through to Conon Street.

Leaves Post Office.	Arrives Gladstone.	Arrives Waikiwi.	Leaves Waikiwi.	Leaves Gladstone.	Arrives Post Office.
A.M.	A.M.	A.M.	A.M.	A.M.	A.M.
7 30	7 45			7 45	8 0
8 0	8 15			8 15	8 25
8 25	8 35	8 40	8 40	8 45	8 55
9 5	9 15			9 15	9 30
9 30	9 45			9 45	10 0
10 0	10 15	10 25	10 25	10 30	10 35
10 35	10 45			10 45	11 0
12 5	12 15			12 15	12 30
12 30	12 45			12 45	1 0
1 0	1 15	1 20	1 50	1 55	2 0
2 30	2 45			2 45	2 55
3 0	3 15	3 20	3 21	3 25	3 30
3 30	3 45			3 45	4 0
4 0	4 15			4 15	4 30
4 30	4 45			4 45	5 0
5 5	5 15	5 20	5 30	5 40	6 0
5 35	5 50			5 50	6 0
6 5	6 15	6 20	6 25	6 35	6 50
7 0	7 15			7 15	7 30
7 30	7 45			7 45	8 0
8 0	8 15	8 20	8 20	8 25	8 35
8 35	8 45			8 45	9 0
9 0	9 15			9 15	9 30
SATURDAYS EXTRA					
2 0	2 15			2 15	2 30
9 30	9 45	10 0	10 0	10 15	10 30
10 0	10 15			10 15	10 25
10 30	10 45	11 0	11 0	11 5	11 20
11 0	11 15			11 15	11 30
SUNDAYS ONLY					
10 0	10 15	10 25	10 30	10 45	11 0
12 30	12 45	1 0	1 5	1 10	1 25
1 30	1 45	2 0	2 0	2 5	2 20
2 30	2 45	3 0	3 0	3 5	3 20
3 30	3 45	4 0	4 15	4 20	4 30
4 30	4 45	5 0	5 0	5 5	5 20
5 30	5 45	6 0	6 0	6 5	6 20
6 30	6 45	7 0	7 0	7 5	7 20
8 10	8 20	8 25	8 25	8 30	8 40

☞ Until further notice the following extra Trams leave Post Office for Gladstone :—8.30 a.m., 1.25 p.m., and 2.5 p.m. Leave Gladstone for Post Office : 8.40 a.m., 1.40 p.m., and 2.15 p.m.

Leaves Post Office.	Arrives Biggar Street.	Leaves Biggar Street.	Arrives East Road.	Leaves East Road.	Arrives Post Office
A.M.	A.M.	A.M.	A.M.	A.M.	A.M.
7 30	7 45	7 45			8 0
8 5			8 15	8 15	8 25
8 30	8 45	8 45			9 0
9 0	9 15	9 15			9 30
9 30			9 40	9 40	9 50
9 30	9 45	9 45			10 0
11 30			11 40	11 45	12 0
12 5	12 20	12 20			12 35
12 35	12 50	12 50			1 0
1 5	1 20	1 45			1 55
3 0	3 15	3 15			3 30
3 30			3 40	3 45	3 55
5 10	5 25	5 30			5 35
5 35	5 45	5 50			6 10
6 10	6 25	6 25			6 40
6 45	7 0	7 0			7 10
SATURDAYS EXTRA.					
2 0	2 15	2 30			2 45
6 40	6 55	7 0			7 15
7 25	7 40	7 45			8 0
8 5	8 20	8 25			8 40
9 0	9 15	9 15			9 30
9 30	9 45	9 45			10 0
10 0	10 15	10 15			10 30
11 0	11 15	11 15			11 30

NOTICE.—This Time-Table will be adhered to as closely as possible ; but the Company will not be responsible for any delay of cars arriving or departing at any Terminus.

Monthly ticket issued at the Company's Registered Office,

HENRY E. WHITFIELD,
Traffic Manager.
J. HATCH, MANAGER,
Dee Street.

The timetable published by the Invercargill & Suburban Tramway Company in January 1887. When this company took control on 1 October 1886, one of the directors of the company was J.G. Ward (later Sir Joseph), who was later to serve as Prime Minister of New Zealand. *Southland Museum*

Southland Tramway Company Limited
Horse-Trams

The *Southland Times* made brief mention of the rolling stock when they reported on the first trial trip to Gladstone on 5 December 1881:

> The cars – made by the Starbuck Company, Birkenhead, are very fair specimens of their class, although not so large as those used elsewhere. They are fitted with conductor's bells, lamps, automatic fare-takers, and other conveniences.

The total number of horse-trams used on the Invercargill lines is not known. Photographs show three different types of single saloon trams with clerestory roofs and platforms at each end. All the cars had saloon sides with curved lower rocker panels and straighc upper waist panels. The first cars had six saloon windows on each side while two photographs feature a much longer car with ten saloon windows on each side. Another had sever. saloon windows.

Above: The city terminus in Dee Street looked so peaceful in the early 1880s as a lone horse- tram waits for departure time. One-horse tram No. 2 proudly displayed the company name on the curved rocker panel. *Burton Brothers*

Below: After studying the number on the front apron, this tram is possibly No. 1, built by the famous British ho:se- tram builders Starbuck and Company of Birkenhead. *Southland Museum*

Above: A three-horse large saloon tram with ten drop-sash saloon windows on each side proceeding down Dee Street at a speed of probably about ten miles per hour heading for Waikiwi. The young lad standing next to the driver wearing a large starched collar seems fascinated with the workings of the horse-power. This view of Dee Street was taken looking south-east from Don Street. *Southland Museum*

Below: The same tram waiting for departure time to Waikiwi at the city terminal in Dee Street, showing a splendid Victorian band rotunda in the foreground.

Above: A seven-window single saloon tram in Dee Street. *Muir & Moodie, Museum of New Zealand Te Papa Tongarewa*

Followng the voluntary liquidation of the Southland Tramway Company in October 1885, a new company, the Invercargill and Suburban Tramway Company, took over a year later, only to face competition from a newly formed company, the Invercargill Omnibus Company. The omnibus company immediately reduced fares, forcing the tramway company to also reduce fares on the Waikiwi line because of the direct bus competition: to Avenal from 2d to 1d, to Gladstone from 3d to 2d, and to Waikiwi from 4d to 3d. Other tram fares were unchanged. It is recorded that it was the tramway company that survived, though not very successfully. J.O.P. Watt in his book *Centenary of Invercargill Municipality 1871 – 1971* went on to say:

Perhaps because it was spurred on by competition, the tramway company tried the use of a double-deck tram ... A newspaper paragraph of 21 June 1887, said:- "The newly appointed car was run for the first time on Saturday afternoon (18 June), when a number of directors and other gentlemen enjoyed a short ride, while comfortably ensconced on the roof of the vehicle. So pleasantly smooth was the journey, and so far above the sound of the jarring wheels were the passengers, that a preference for outside riding in favourable weather was unanimously expressed. A car laden top and bottom will doubtless require more than one horse, but arrangements can easily be made for placing an additional animal in the harness... The necessary work has been designed, and to a large extent executed, by *Mr Whitfield, the company's manager."

The double-decker did not prove to be popular in Invercargill with no protection from those Antarctic winds.

*Mr Whitfield had come from Christchurch where the Canterbury Tramway Company had been operating double-decker horse-trams since April 1882. This car may have been modified from an existing single-decker car by Mr Whitfield.

Below: The same type of horse-tram crossing the Waihopai Bridge at Gladstone on the Waikiwi route. *Muir & Moore, Trevor Terry collection*

INVERCARGILL
Otatara

Steam Tramway

Oreti Tramway Company

Subsidiary of the Otatara Land Proprietary Company Limited
From 25 March 1908 to c1912

Gauge 3 feet 6 inches
The population of Otatara was 192 when the system opened.

The Suburbs and Streets Served by the Steam-Tram

From the estuary side of the Invercargill Railway Station at the end of Tweed Street, across the New River Estuary, Stead Road, Oreti Road, to Dunn's Road, Otatara.

Route length: Approximately 1½ miles. 40 lb rails.

Above: The official opening of the Invercargill to Otatara tramway by the mayor of Invercargill, MrW.B. Scandrett, on 25 March 1908. *Otago Witness, Colin Robertson collection*

Oreti Tramway Company

Steam-Tram

Merryweather (One tram locomotive)
Merryweather No. 2 standard type.
Fleet Name: *Scotia*
Year built: 1878
Imported new by the Wellington Tramway Company in 1878, then purchased by the Dunedin, Peninsula & Ocean Beach Railway Company Limited in December 1880, later used as a shunting locomotive at Woolston, Christchurch between 1902 and 1906, then sold to the Otatara Land Proprietary Company in 1907.
Builder: Messrs Merryweather and Sons, Greenwich, England.
Maker's number: 87.
Specifications as built: Wheel arrangement: 0-4-0, 7 inches x 11 inches inside cylinders. Wheel diameter: 2 feet. Wheel base: 4 feet 6 inches. Water condenser on roof. Tank capacity: 300 gallons. 10 h.p. boiler, rating 120 lbs psi.

Tramway built by the Otatara Land Proprietary Company to give direct access to their residential development at Otatara in 1907. 223 freehold sections of not less than one acre each, residential suburban sections of about three acres each, also one section of eight acres, and another of 32½ acres were offered for sale. The development company offered a free ticket to people who bought sections. "Any purchaser who erects within twelve months of the date of sale on the land bought by him, an approved dwelling house, will be entitled to a free ticket on the tramway for two years."
The "grand official opening" took place on Wednesday, 25 March 1908 with about 1000 people in attendance. A luncheon for official guests was held in a marquee and other tents provided tea and cakes and other "light edibles".
The *Southland Times* reported after the first week of operation:

> The tram to Otatara has been running a week now, and during that time and not withstanding unfavourable weather, over 500 passengers have been carried. Those who walked from the terminus to the Oreti River, about one mile, were much impressed with the view of the fine river which in the near future is to be one of the favourite resorts of the townspeople.

The company had planned to extend the tramway to the New River when 15 per cent of the land had been sold. The tramway and the development were not a commercial success and the tram ceased running within a few years. The tramway company had endeavoured to get the Invercargill Borough to take over the tramway and extend it to the New River within two years and later to the beach. Councillors were not interested in taking over an unprofitable operation.
From September 1910 the tram locomotive was used on construction work on the New River Estuary reclamation, Invercargill, until 1921, then on the PWD Orawia Railway extension of the Tuatapere branch. Scrapped 1925.

The *Auckland Weekly News* of 21 November 1907 published this photograph with the following caption: "Opening of a new southern pleasure resort. The steam tramway connecting the city of Invercargill with the Otatara grounds opened to the public recently." The photograph shows the Merryweather tram locomotive and the two former Wellington horse-trams. The leading tram had been renumbered No. 2. The trams had had the end canopies rebuilt with glass fronts. *E.Denton, Auckland Weekly News*

The ex-Wellington Merryweather steam-tram and trailer – renumbered No. 1 by the Oreti Tramway Company, at the Otatara Station in Oreti Road. *Courtesy Marilyn Bunce*

After the public passenger tramway service had closed – the Merryweather tram locomotive and one tramcar with a rake of trucks being used for clearing work at the end of Bryson Road. *Russell Grigg collection*

BIBLIOGRAPHY

Books

Bunce, Marilyn, *Sawdust and Scholars*, Otatara School Centennial Committee, 1979.
Campbell, Bill, and Hargreaves, Ray, *Straphangers and Grippers*, Otago Heritage Books, 1994.
Dangerfield, J.A., *Dunedin's Matchbox Railway*, New Zealand Railway & Locomotive Society, 1986.
Grant, R.D., and Dangerfield, J.A., *Steam and Horse Tram Working in Dunedin 1879-1905*, published privately, 1991.
Johnston, Mike, *High Hopes*, Nikau Press, 1987.
Johnston, Mike, *Nelson's First Railway and The City Bus*, Nikau Press, 1996.
On the Move: Christchurch Transport through the Years, 3 Rails in the Roads, Christchurch Transport Board & Tramway Historical Society, 1990.
Stewart, Graham, *The End of the Penny Section*, A.H. & A.W. Reed, 1973, Grantham House, 1993.
Stewart, Graham, *When Trams were Trumps in New Zealand*, Grantham House, 1985.
Stewart, Graham, *Always a Tram in Sight*, Grantham House, 1996.
Walsh, T.W., *Devonport and the Old North Shore*, 1924.

Periodicals

NZ Railway Observer, NZ Railway and Locomotive Society, Wellington.
Tramway Topics, Wellington Tramway Museum, Tramway Historical Society, Tramway Division, Museum of Transport and Technology.

Newspapers

New Zealand Herald, Auckland.
Weekly News, Auckland.
Thames Star, Thames.
Ohinemutu Gazette, Paeroa.
WanganuiHerald, Wanganui.
Wanganui Chronicle, Wanganui.
Dominion, Wellington.
Evening Post, Wellington.
Press, Christchurch.
Otago Daily Times, Dunedin.
Evening Star, Dunedin.
Southland Times, Invercargill.

Other Sources

Gerald Ditchfield research notes.
Ron Grant research notes.
David Hinman research notes.
Dayle Maffey research notes.
General Assembly Library, Wellington.
National Archives, Wellington.

NEW ZEALAND HORSE-TRAMS
Their Origins & Construction –
The Horses and Travel in the 19th Century

The first horse-drawn tram service in the world started in New York in November 1832; Paris was the first European city to build a horse-tram line in 1855. Another five years were to pass before the first horse-drawn tramway opened in England; the necessary Parliamentary approvals had caused years of delay. In August 1860 the first true British street tramway for regular passenger service using horse-drawn trams on one-and-a-half miles of track opened at Birkenhead. It had been built by an American with the believe it or not surname of Train (George Francis Train).

Above: The first horse-tram in the world, built by John Stephenson of New York in 1832. *Right:* The opening of the first horse-tramway in England at Birkenhead.

It seems rather remarkable that only 21 months later a one-horse-drawn tram started a passenger service in the far-off colony of New Zealand between the City of Nelson and its port in May 1862. The tram had been built by a Sydney coach builder who the previous year had built the first horse-drawn tram vehicles for service in Sydney when a line opened in Pitt Street.

The discovery of gold on the West Coast of the South Island saw the construction of many horse-drawn lines with timber being used for rails during the 1860s. Although steam-powered tramways inaugurated services in Wellington, Dunedin and Christchurch, where they continued to serve in conjunction with horse-drawn services, by the 1880s it was found that horse-drawn trams were cheaper to run for the small volume of patronage offering and were preferred by the public and were discarded by Wellington and Dunedin.

Horse-tram construction was a specialised business with the American companies, John Stephenson Company of New York, who had built the pioneer New York horse-tram, and J.G. Brill and Company, Philadelphia, also an early manufacturer for the world market, seemingly ahead of English and Continental builders in the 1880s. Construction had to be light for the sake of the horses and it is recorded that American woods gave the required strength with less weight than European timber.

Mr John Stephenson boasted that the quality of American woods had much to do with his success, the selection and preparation of materials requiring three to four years. The timbers used were white oak, white ash, poplar, basswood, hickory, beech, maple and pine. The English used mainly teak imported from the West Indies which was a heavy ship-building timber.

New Zealand companies imported American, English and European built horse-trams, and as local skills developed, orders were placed with New Zealand coach builders who often copied proven American and English designs.

Manufacturers advertised: "Seats and backs made of 'fancy' wood, cushioned seats cost extra. 'Super Spring Gear' which excels in comfort to passengers, ease to horses, and economy of maintenance; ventilating ceiling, always operative for good air; life guards before the wheels, remove obstacles and protect life; lantern top has illuminated glass lettering, conspicuous at night. One horse can operate these cars, the largest seating 22 passengers, on level roads, and the smallest for only 10 passengers on grades equal to 6 per cent."

The services in Wellington, Christchurch and Invercargill were operated over flat terrain. Dunedin had to use leader horses on services from Cargill's Monument up the rise in Princes Street to the Octagon. Auckland must have been a nightmare for horse traction, with all services having to climb either Wellesley Street West or Wellesley Street East. Tests found that three horses could take a tram containing 35 passengers up the Wellesley Street West grade of 1 in 12.

Clydesdale leader horses with a 'hill boy' astride were used to assist each tram on the Wellesley Streets grades. A 'hill boy' would wait at the bottom of the hill, hook a chain to the pole of the tram and assist the lighter tram horses on the uphill climb, then ride down the grade and wait for the next tram.

Above: A John Stephenson horse-tram from the manufacturer's catalogue of the 1880s. *John Stevens*

Above: A double-decker horse-tram from the J.G. Brill catalogue of the 1890s. *John Stevens*

The Horse Power

The trams were served by robust animals that were changed three times a day. But the human material was held to be more durable, and the same driver had to stay on duty throughout the three-horse shifts. In Auckland horses were required to walk almost 30 kilometres a day, getting a quarter-hour spell between each trip. Hauling was a fairly heavy task, particularly when the 3-ton trams were overcrowded. An average of 200 horses were stabled at each depot. The stalls were divided, Yankee fashion, by hanging bars and at times there was great commotion on the cobbled floor when a number of horses, forgetful of tomorrow's toil, would playfully push the bars about. A small stocky type of horse proved the best for the job. For the Auckland services large numbers were brought from Wanganui. They were a fine upstanding lot, all possessing good blood and breeding. Later they were supplanted by a breed from the Waikato, until the late 1890s, when suitable horses were imported from Queensland for £l a head, plus £1 for freight. These fiery untamed animals were shipped in droves and swum ashore at St Hellers to be broken in by the company on their 160-hectare farm. Many a beast bred on the ranges of Waikato and Queensland settled down to a career of steady toil in front of a tram.

Special departments at the depots prepared the food for the hungry horses. Most companies grew progressive crops to keep up a steady supply of maize, oats and chaff on out-of-town farms. The daily ration for one horse consisted of 4.5 kilograms of oats, 1.3 kilograms of maize, half a kilogram of beans (weighed after crushing), just over 6 kilograms of chaff and 2 kilograms of hay, with I kilogram of bran, or all equivalent of green food. Carrots were also included.

Harness was always a costly item, having originally been imported from the United States, but local harness-makers later fulfilled orders. Each groom had a team of horses, ranging from 15 to 20, under his care to feed and keep tidy with glossy coats and manes brushed. Grooms oiled the many sets of harness, made reins pliable and kept collars and traces to a high standard. In the smiths' department stacks of iron shoes for the steeds and iron tyres for the buses reflected the glow from the forge fires. The health and general well-being of the animals were always foremost. A small horse hospital was part of a depot, and veterinary attention was considered as important as maintenance expenditure on the trams. Big money was invested in the horses, each car requiring between seven and eight animals to work the shifts.The lads who brought the horses in from the depots and took the tired horses back to the stables were called "change boys".

*In Dunedin the grooms at the stables had 17 horses to look after and were responsible for their grooming, the horse boxes and for bedding them down, as the horses were stabled night and day. Each groom worked 77 hours a week, starting at 5.30 am and working until 5 pm or 8 pm on alternate nights. They were given every second Sunday off, after they had cleaned their 17 horses and bedded them down. There were four stables, Cumberland Street, The Grove, Normanby, Ocean Beach and David Street, Caversham. Horse-trams were housed at each stable where they were washed and greased each night. Everything had to be well cleaned, the floors of the stables and the horses were spotless. The night watchman at each depot had to wash and oil all the harnesses, then at about 3 am, feed about 100 horses. The feed

*From an interview with William Webb at the age of 84 years who worked on the Dunedin horse-trams. Recorded by railway historian Jim Dangerfield in July 1965.

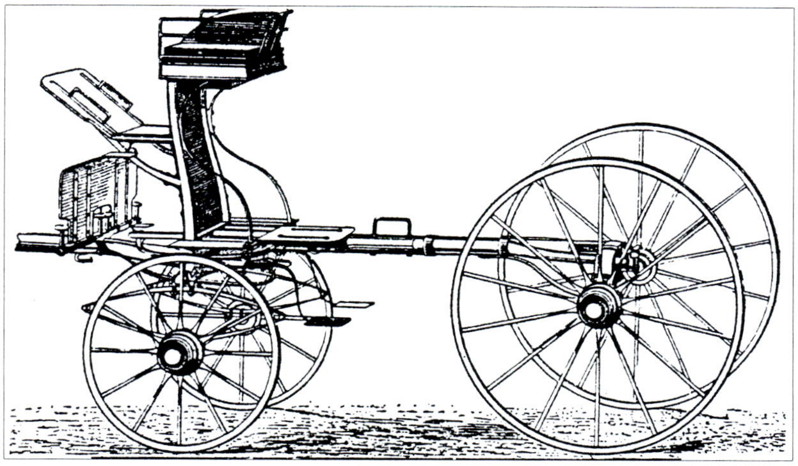

A Brake – a four-wheeled vehicle used in "breaking" horses. A decoy horse (a trained horse) was placed on the other side of the centre pole to show the new horse what to do.

already mixed – chaff, oats, wheat, barley and bran – came down a chute into a bag, which the watchman would carry around until each horse was given his "whack of feed". Unbroken horses were taken to Ocean Beach and broken into harness on the sands, then driven into the blacksmith's shop at the Cumberland Street stables to be shod. They were yoked to a Brake, a four-wheeled vehicle with a high seat for the driver and a decoy horse (a trained horse) was placed on the other side of the centre pole to show the new horse what to do.

One of the leader horse boys was called "Wingy", as he had lost half an arm in an accident. The horses at times really had hard work starting the trams. The big pull from High Street up to the Octagon meant they had to start at a gallop to get to the top. A leader horse was always attached to the double-deckers and the large single-deck cars, with a long rope between the three horses attached to the tram. Mr Webb, who was employed as an extra leader rider recalled: "my horse was hooked on to the side of the decker tram. It was a dangerous place for if your horse was to fall the driver could'nt see you as passengers were hanging on where they could. A lot got a free ride as the conductor couldn't collect all the fares in one or two streets. But there was a lot of honest people who would leave their pennies for the conductor. There was only one bad accident on the horse-trams in my time when a new conductor was warned not to throw the traces on a pony called 'Jimmy the Rat'. He remembered for two or three trips and then forgot. Sadly 'Jimmy's' two hind feet hit him and he was killed. There was another accident in Princes Street South when a horse tripped and fell before the tram could stop – and the tram ran on to the horse and ripped open its stomach and a lot of its entrails came out. There was a hold-up until the vet came and sent for a bucket of milk, washed the entrails , put them back and sowed up the outside skin. The horse was put in a nearby paddock and looked after. He was back at work in a couple of months. Some horses would slip on the tram rail when starting a tram and cut their knees, and they would be put in the horse hospital until they were ready to work again. The centre horse had to do the biggest pull as he was between the rails. Mr Webb recalled the names of some of the horses, there was 'The Camel' who stood 18 hands high; 'Man Eater' earned his title for his bad habits; 'Stilts' was an awkwardly built animal; 'Slummocky', 'Bessy', 'Black Tom', they all had their own personalities. Mr Webb recalled that after the horses were sold, it was difficult to get any work out of them until a bell was rung as a signal to start even if the work was only pulling a plough."

Leaning into the collars with ears pricked forward, necks and knees bent, a team would settle into position like a football scrum and appear just as keen. By habit the horses followed the centre line of the tracks and kept the car on the rails. Not only was it their duty to pull the car, they had to act as a brake on the down grades when the handbrakes were faulty. Large advertising boards added excessive weight and many an unnecessary burden must have been imposed on the hapless horses. The blood of the animals which provided propulsion for a city in those years was the life blood of the city itself.

Travel in the 1890s

**The horse-tram drivers knew nearly everyone who travelled and should a regular passenger be late over breakfast, the driver occasionally would wait. Schedules were not strictly followed and on the rear platform the conductor would be watching for his regular clientele. In addition to halts at important traffic points, the trams would stop when signalled by the wave of a hand, umbrella or walking stick. It was general practice when the car approached to walk out onto the street and step aboard, the horses slowing momentarily before resuming their 10-kilometre-an-hour pace. Naturally, when a woman or an elderly person required transit, the driver would call the horses to a halt. With the shrill blast of the conductor's whistle and the cheery "Giddy up" as the driver flicked a whip over the shiny coats of the oats-charged horses, the tram would surge forward and the clip-clop of the horses' hoofs would echo back through the saloon.

On the front platform, exposed to the weather, sat the driver, who had only a small splashboard to shield his legs. His hands firmly held the ribbons that gave him complete command. There were no "Do not talk to the driver" platform notices to discourage male passengers from exchanging racing tips with the driver as they boarded the car. On entering the small passenger saloon, which had a smell peculiar to the horse-car – heavily scented with kerosene lamps and sweating horses – passengers would sit on one of the two wooden lengthwise benches. The conductor would issue the passenger with a ticket the size of a postage stamp, punch it with a bell-punch for all fellow patrons to hear and then one could relax and enjoy this latest form of suburban travel.

The rails made the ride so smooth that reading was no effort, and the males would spend the journey with their heads buried in the morning paper. Because of the relatively low resistance offered by iron wheels on iron rails, the cars could handle bigger loads than the horse-buses. Smoking was definitely not permitted in the presence of women, the back platform being the accepted place for men to enjoy their pipes. From this early ruling must have stemmed the reason for men persisting, through the ensuing years, in gathering on tram platforms, blocking all entrances. Frisky young mares drawing spring carts and placid wagon teams challenged the right of the road at intersections. After thumping over a crossing with another tramline the car would glide like magic around a curve with the ease of a roller-coaster. At the city terminal the men always doffed their hats to the ladies seated opposite, and then everyone alighted fresh from this new mode of travel. The driver in his weathered clothes unhitched the team and, leaning back on the heavy swingle-tree, piloted the horses to the other end of the tram with trace chains dragging in the dust, while the passengers briskly made their way past a line of hansom cabs to their places of employment.

**"Travel in the 1890s"; the author researched contempory newspapers and interviewed his maternal grandmother who was born in 1875, and as a young lady living in Auckland, travelled daily by horse-tram from the suburb of Newmarket into Auckland city in the 1890s. She died in 1977 in her 102nd year.

NEW ZEALAND STEAM-TRAMS AND THEIR ORIGINS

Ron Grant

Ron Grant is a retired civil engineer, a former District Commissioner of Works, Ministry of Works. He enjoys New Zealand history, particularly the 1850-1910 years of infrastructure and industrial development. He has contributed technical articles to New Zealand and international historical magazines since the 1950s. He is a former member of the Institution of Professional Engineers, New Zealand National Engineering Heritage Committee, and remains an active member of the Canterbury Chapter of that body.

If one believed the claims of Merryweather and Sons Ltd, the noted steam and fire engine builders of Greenwich, London, one would accept that the Wellington Tramway Company's system which opened on August 24, 1878, and which employed Merryweather steam-tram engines, was the first steam tramway in the Southern Hemisphere. Merryweather's publicity brochures were often highpowered and not always factual. The Pernambuco Tramway Co, Brazil, served the city now known as Recife from 1867 and used Manning Wardle's tramway locomotives. Pernambuco is a good 8 degrees south of the line, and for that matter there were other Southern Hemisphere steam tramways that antedated that of Wellington. Even in New Zealand the Wellington Tramway Company was not the first.

Chaplin's 12-horsepower vertical boiler locomotive. The Grahamstown & Tararu locomotive was identical except that it had spoked instead of solid wheels. *The Engineer, 1875*

That honour belongs to the Grahamstown and Tararu Tramway Co of Thames which opened on 28 December 1871. From 1869 the Auckland Provincial Government had built a number of horse and gravity tramways connecting the gold mines with the ore-crushing batteries along the sea front. Ore for treatment and firewood to feed the hungry boilers of the batteries was the freight, and in general, passengers were not carried. The new tramway served both the township and at each of its extremities, the wharves from which steamer service to and from Auckland sailed. Both freight and passengers were catered for. Indeed, some of the scheduled passenger runs were mixed with a short rake of ore or merchandise wagons in the rear of the passenger cars. The company's line was connected with the Provincial Council's tramways, thus serving several mines and batteries.

There were two locomotives, the first being a second-hand Chaplin-geared vertical boilered locomotive (believed to be No 1182 of 1870), first used by the Kawa Kawa Coal Company but discarded by that company as having insufficient power for the traffic offering. Being a standard gauge (4 ft 8½ inch) locomotive it had to be altered to the Tararu Company's 3 ft 6 inch gauge by the Auckland firm of Fraser and Tinne. Delivered to Thames on 2 December 1871, it went to work immediately. The Chaplin locomotive was an inexpensive design for light intermittent shunting duties. The route was level and speeds were low largely because of track conditions.

There is conflicting evidence concerning the second locomotive which was built by Fraser and Tinne. I believe it to have been a vertical boiler machine practically identical with the Chaplin, the first of the Chaplin "look alikes" that the firm constructed in the 1870s for contracting and industrial use. Mr Fraser had used the Kawa Kawa locomotive as a pattern for his future production. It was placed on the line on 23 December 1872.

In contrast, the Wellington Tramway Company's operation was a normal steam passenger tramway serving the central part of the city. Merryweather and Sons had been building tramway engines since 1875. In Britain, because of the very restrictive tramway legislation obtaining until 1879-1880, horse traction was the only practical method of operation. Consequently, all Merryweather's production went to tramways on the Continent of Europe and a little later to Empire destinations. In spite of this the firm developed a successful series of standard designs ranging from the 4-ton engine with 6-inch diameter by 9-inch stroke cylinders, eventually to a 9½ ton, 10 inch by 14 inch machine.

The Wellington engines were the standard 6½ ton, 7 inch by 11 inch model with a small but quick

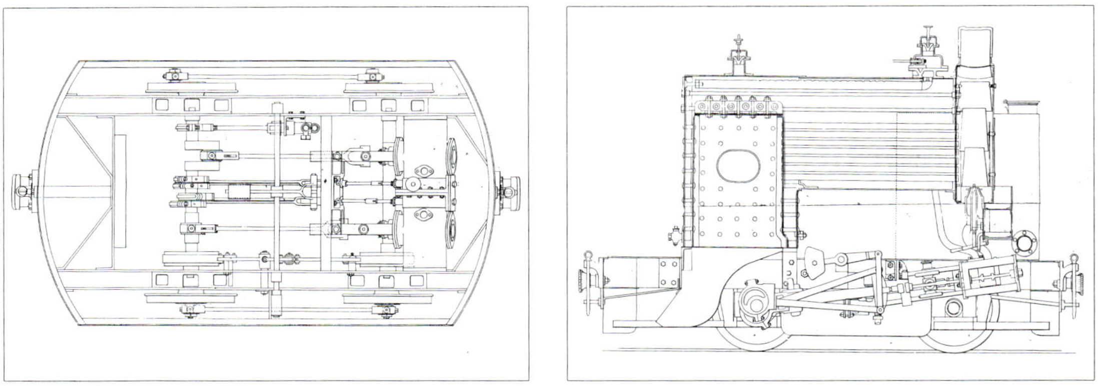

Sectional elevation and machinery layout plan of Merryweather's 7 inch by 11 inch standard tram engine as supplied to the Wellington Tramway Company. *Tramways, D.K. Clark, 1881*

steaming locomotive-type boiler enclosed in an all-over cab made like a short tramway car, having several windows. This disguise was supposed to calm horses which might otherwise take fright at the live snorting steam engine it concealed. The two cylinders were horizontal inside the framing and drove directly to cranks on the driving axle. Coke was the fuel to obviate any smoke nuisance. A 300-gallon tank to hold a supply of cold water for condensing the steam was placed on the roof. Steam also must not be seen. While a normal blast pipe was fitted, most of the exhaust steam was condensed and passed to the tank until the temperature of the water in the tank approached boiling point when the hot water would be discharged into the city stormwater system and a fresh supply taken. The 7-inch model was expected to draw two loaded cars weighing 14 tons up an incline of 1 in 30.

There were eight engines, Merryweathers Nos 60 to 64 and 85 to 87 of 1877-1878, all of which bore names. While the locomotives worked well, the same could not be said of the track which was of a light grooved strap-rail form, wholly unsuitable for a locomotive line. It was not long before the rails were crippled, bent and twisted out of shape. Derailments became a frequent occurrence. Much of the track was relaid in steel rails and the locomotives were set aside in favour of horse traction. Being practically new, the locomotives were quickly taken up by light railway and bush tramway interests. The Kauri Timber Co had three of them. Most of them had quite long lives on rails a long way from Wellington.

The Dunedin Tramways opened on July 7, 1879. This was a private venture by David Proudfoot, a civil engineering contractor. Early in 1878, Hugh A. Macneil, of the local importing firm of Arthur Briscoe and Co, travelled overseas empowered to order rails, plant, locomotives and rolling stock. His instructions were to procure the best equipment that was available in Europe and America.

Macneil visited the 1878 Paris Exhibition where a number of British and Continental-built tramway locomotives competed for the coveted Exhibition medals. The winner of the section was the English firm of Henry Hughes and Co, their exhibit being destined for the Southern Tramways of Paris. Merryweather and Sons, up to that time the leading tramway locomotive manufacturer, received only second prize for an

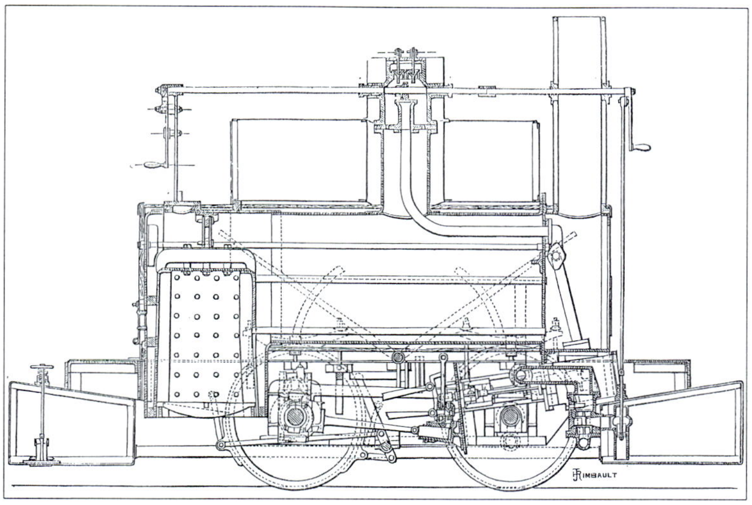

Sectional elevation of Hughes locomotive for the Southern Tramways of Paris, the engine that impressed Macneil as suitable for Dunedin. Controls are at each end, while cylinders are inside the frames on these standard gauge engines. *Tramways. D.K. Clark, 1881*

engine similar to those exported to Wellington that year. The upshot was that four 3 ft 6 inch gauge versions of the Exhibition prizewinner were ordered from Hughes' Falcon Works, Loughborough.

Hughes, a builder of contractors and industrial type locomotives, had been developing his tramway locomotive, an adaption of an industrial type saddle tank since 1876. Like Merryweather he was up against the legislation but, in spite of this, arranged trials of his engines in several English cities. The legislators remained unconvinced, and while a number of Hughes engines were running on the Continent and due to a quirk of the legislation, at Govan near Glasgow, Hughes eventually could not stand the financial outlay and lost control of his company. He emigrated in poor circumstances to New Zealand; later he founded a most respected patent agency business, for many years the leading organisation of its kind in Australasia.

The four locomotives, Hughes Nos 326 to 329 arrived in Dunedin in March 1879. They were conventional four-coupled outside-cylinder saddle-tank engines enclosed within a wooden cab. All controls were brought to the right hand side of the horizontal boiler where the sole engineman had a fair view ahead. Firing was done at each terminus, coke being the fuel, while the exhaust steam passed through the condenser tubes mounted on the cab sides. Condensate then collected in low level tanks at each end of the locomotive. This was not recirculated into the boiler but was discharged to waste into a specially built sump at Cargill's Monument, the city terminus, while a fresh saddle tankful of feedwater was taken on board. While the engines were quiet in operation, the condenser caused back pressure and absorbed a great deal of power. Following the depot fire of December 1880, the condensers on the three surviving engines were disconnected with a corresponding increase in "bark" and pulling power. Designed to run at a normal service speed of 15 miles per hour, by regulation their speed was not to exceed 10 miles per hour within the city.

Macneil travelled on to Baldwin's Works in Philadelphia, USA, where he purchased a second-hand combined engine and car (a steam-tram car) recently repossessed from the financially troubled Galveston, Brazos and Colorado narrow gauge railroad and a new vertical boilered steam motor. The steam tramway engine, or steam dummy as it was known in America, had developed on similar lines concurrently with that in Europe. Several vertical boilered steam tramcars and some small dummy locomotives had been built by various makers, but by and large the horse was still unchallenged in America.

Baldwin entered the field with a vertical boilered inside cylinder steam streetcar which commenced running in Brooklyn on Xmas Day 1875. In June 1876 it was transferred to the Market Street Railway of Philadelphia carrying passengers to the Centennial International Exhibition – just the latest thing! At the close of the exhibition after a total of a year's running, it was brought back to the works and completely rebuilt with an iron frame independent of the car body and outside cylinders. This, the prototype Baldwin steam car, was the design for some 12 built over the following three years. The sixth of these, construction number 4139 built in August 1877, went to Galveston, but along with another locomotive that the company could not pay for, was repossessed and caught the eye of Macneil when he visited Baldwin's Works. After regauging from three feet to three feet six inches, "Washington" as it was named, was despatched to Dunedin on the "Sonntag" leaving New York on 30 April 1879.

Washington went into service on the Castle Street route on 19 September departing from the Monument every half hour between 7.00 am and 9.00 am. Its driver sat at the controls right in front of the car ahead of the boiler which he also fired. Immediately behind the boiler compartment was the 20-seat passenger saloon which was entered from the rear platform. The whole was mounted on 30-inch diameter four-coupled wheels with a wheelbase of 6 ft 6 inches and was capable of sixteen to eighteen miles

The prototype Baldwin steam car of 1876 – practically identical to "Washington", that ran on Dunedin's streets for only five weeks. *The Engineer, 1877*

per hour. The coachwork and fittings were made by the J G Brill Co, the first of the many Brill cars that came to New Zealand.

There was no means of turning the car either at the Monument or at the Water of Leith. The service left the city with the driver facing forward and ran in reverse as it returned to the city. The conductor not only collected fares, but if he could, kept a forward lookout from the platform, signalling the driver in the best way he could. Sooner or later an accident was bound to happen. It came on Friday 24 November during the lunchtime rush.

A drayman was unloading cement barrels when his horse shied at the approach of Washington. He rushed to the head of his horse but was struck down, run over and was so seriously injured that he died that evening. The City Council quickly ruled that if Washington was to continue to run, it must be with the driving end leading at all times. Washington was thus set aside after only five weeks in traffic.

Soon afterwards it was completely rebuilt. The engine portion, with the boiler centrally repositioned and with an ornate locally made all over cab, reappeared on the streets in March 1880 as steam motor "Grand Pacific". The car body work was mounted on a leaf sprung railway type underframe and continued as a trailer car still bearing the name "Washington".

Macneil's other American purchase, "Atlantic" (No 4522 and ninth of her class), was one of the last Baldwin-built vertical-boilered steam dummys. She was fast and 60% more powerful than the Hughes' engines. The four-coupled wheels were driven by 9-inch by 12-inch horizontal cylinders.

"Atlantic" did not comply with British Board of Trade requirements. There was no condenser so that steam and smoke escaped both visibly and loudly to the atmosphere, the side rods clanged and to make matters worse, the steam whistle had a most mournful and penetrating note. American steam dummys were expected to warn all of their presence, rather than slink by almost silently and invisibly, as was demanded of steam motors in Great Britain. The City Council was quick to notice, or had brought to its notice, that great operating difference.

The deliberate arson of the main tramway depot in December 1880, for which no one was ever brought to book, destroyed or severely damaged the whole of the locomotives and rolling stock except for two cars stabled at Ocean Beach. "Atlantic" was the first motor restored to service and renamed "Gordon" after the British military hero. "Howling Gordon" continued to provoke city councillors until the end of steam traction in Dunedin in 1884.

Baldwins continued to develop the steam motor for street railway service. The limited steam raising capacity of the vertical boiler was only too apparent for any improvement beyond engines of Atlantic's size and capacity and a quick steaming locomotive-type boiler was designed. More powerful and faster models were added to the range and by 1880 107 steam motors and twelve steam cars had been turned out for service not only at home but for Central and South America, New Zealand and Australia. Indeed, the New South Wales Government Tramways were to employ 100 Baldwins along with 22 locally built copies (as well as those of other makers), one of the largest steam tramways ever. The eventual output of the firm was some 500 tramways engines in all.

There were two further engines on Proudfoot's Dunedin tramway, both built by Kitson and Co. The first, Kitson tramway list, No 7 of 1879, the prototype Kitson tramway engine and what would seem to be its 1880 fire replacement for which the builder's number is not known for certain (possibly T53, 54 or 55) was supplied to Dunedin in 1882.

Kitson's steam-tram locomotives were regarded as the pre-eminent British engines in their field over the final two decades of the nineteenth century before the electric tram chased steam off the streets. Much of the developmental phase of this type was involved with the New Zealand engines.

The engines had their origins in W. R. Rowan's steam rail cars which were intended for light railway or street tramway service. Its initial form was a light bogie passenger car or goods van with a vertical boiler, engine, driver's position and controls all mounted on the leading bogie which was free to swivel within the car body. Rowan, an English engineer was the proprietor of the Scandia Works in Randers, Denmark, a railway carriage building concern. Merryweather built the steam power bogie and boiler for the initial Scandia built Rowan steam car of 1875. A larger double decked car followed in August 1876, the boiler and steam bogie being built by Kitson and Co and this commenced the separate Kitson list of "Tramway Engines", the so called "T list", being serial number 1.

In 1876-1878 James Kitson built three experimental vertical-boilered tram engines which were mechanically identical to the Rowan power bogies with vertical cylinders mounted on each side of the boiler, the whole being enclosed in a "house" with Rowan's air-cooled condenser mounted on the roof. Rowan was not at all happy with this piracy of his ideas, and while a few other Rowan steam bogies went through Kitson's works, eventually this work went to continental locomotive works.

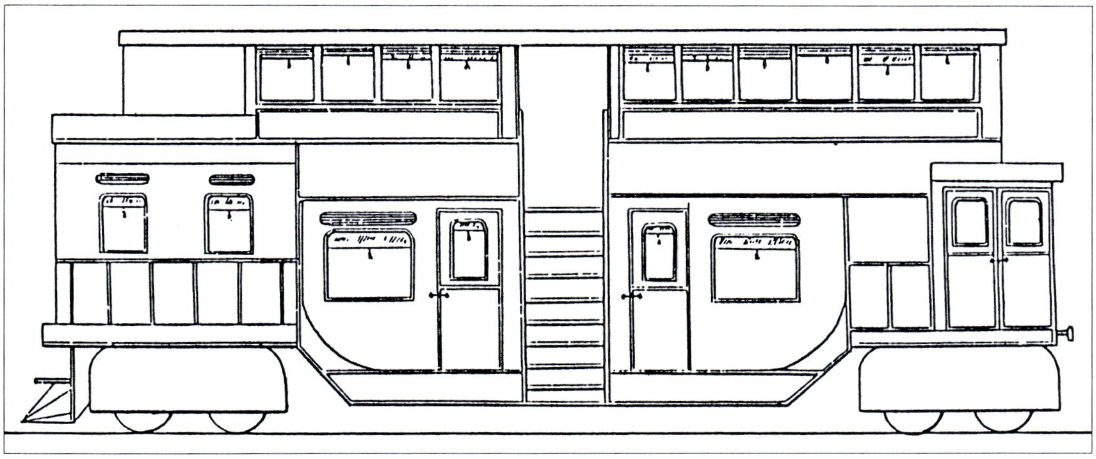

An elaborate Rowan steam car as offered in Captain Rowan's promotional material. Could it have been considered for the Sumner line? The steam power bogie is at the cowcatcher end. *Light Country Railways & Street Tramways. 1878*

Meantime, two personages who were to have much influence on the future of the Kitson tram, appeared on the scene. The Hon. Richard C. Parsons, brother of Charles A. Parsons of steam turbine fame, was on the Kitson staff. The other was Captain Frederick C. Rowan, brother of W. R. Rowan, an ex-soldier who having been seriously wounded in the Maori wars, had set up in practice as a civil engineer and also Kitson's agent in Melbourne.

Captain Rowan compiled a pamphlet "Light Country Railways and Street Tramways", a compendium covering drawings, details of trials and advocacy of his brother's "steam cars" for both city street and light country branch-line service. Also included was information on the trials of James Kitson's "separate engines" of 1876-1878 and the particular advantages of such motive power. It was circulated throughout Australasia.

Knowing that there were proposals afoot for tramways in Dunedin and Christchurch, Rowan visited New Zealand in August-September 1878. The "New Zealand Railway Observer" No 195 – Spring 1988 – gives a full account of the origins of the Kitson steam-tram engines for New Zealand. The following merely summarizes the main events.

David Proudfoot of Dunedin required further engines to work route extensions to South Dunedin and North East Valley which lay beyond the original city concession. One engine was ordered on a trial basis. Rowan's discussions with the Canterbury Tramway Company's directors were most fruitful as that company's orders were then being placed for all locomotives, rolling stock, rails and plant required. The newspapers reported that five "separate engines" were ordered suggesting that the directors had considered Rowan cars and separate locomotives before coming to a decision.

Meantime, Parsons was completely redesigning both the Rowan car power bogie and the Kitson tram engine. The power bogie was to have conventionally mounted near horizontal cylinders and a larger boiler. Two Rowan cars were under construction – T5 at F C Rowan's behest, for demonstration purposes at Glenelg, South Australia and T6 which appears to have been Parsons' experimental steam bogie on which changes were made as his thinking evolved. Years later, it was to run in Sydney. The Kitson standard steam-tram evolved with T7, the Dunedin engine. Late in 1878, soon after the order had been placed but before construction had started, a description probably supplied by Captain Rowan, appeared in the Dunedin papers. "The engine has cylinders 6¼ inches in diameter, 12 inches stroke and is fitted with the latest improvements for condensing steam. The condenser is on top of the cab and consists of 200 copper tubes. ....A circular upright boiler provides the motive power ..." In the event, a locomotive-type boiler was substituted. A variation of Walschaerst valve gear was designed, bringing practically all the wearing parts above the footplate and clear of road filth and grit. The regulator and steam brake control was combined in a single operating lever. The engines were completed in order sequence. T7: the 3 ft 6 inch gauge Dunedin engine being completed first in August 1879, while the standard gauge Canterbury engines appeared at two

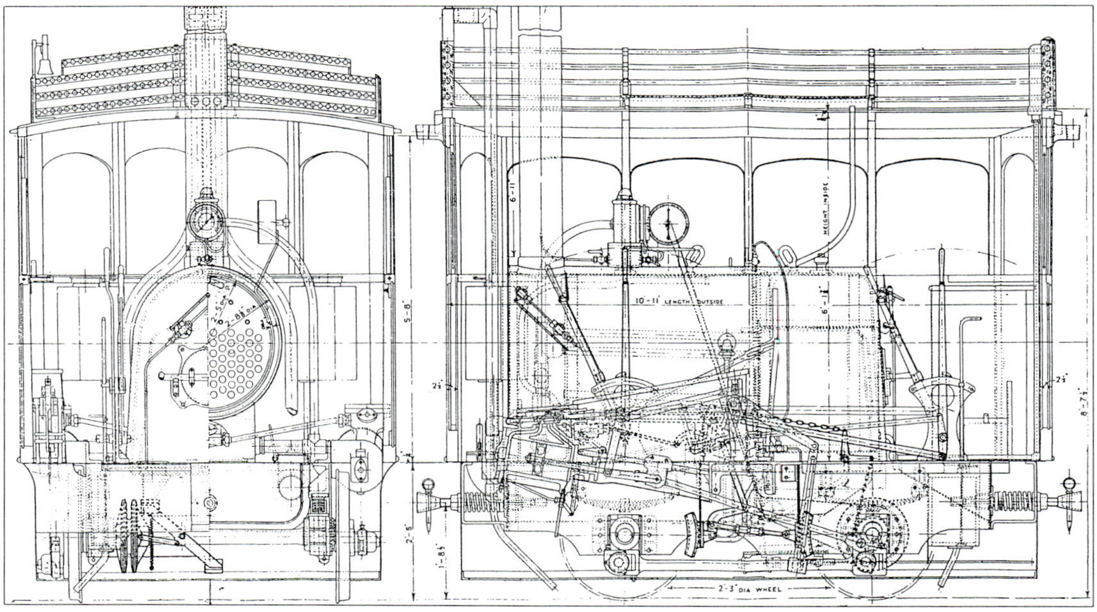

The original drawings for the Christchurch steam trams. The longitudinal condenser arrangement was altered to a transverse layout before completion. *Kitsons of Leeds*

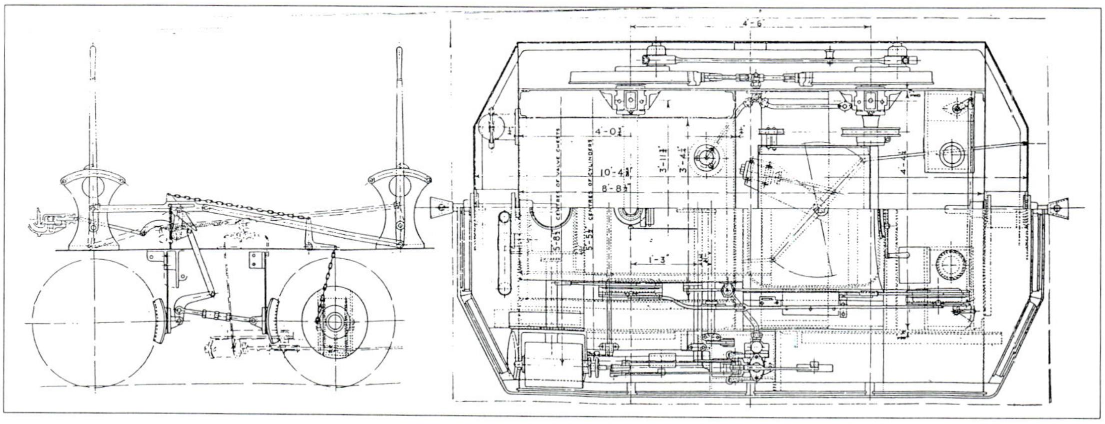

The combined regulator and steam brake control with its single lever mounted at each end of the locomotive, still controls "Kittys" movements at Ferrymead. *Kitsons of Leeds*

to three-weekly intervals over the period August to October 1879. The latter were road tested on the Leeds Tramway Company's lines. A track connection was specially laid in between the Kitson Works and the tramway company's Hunslet Road line. Four engines T8-10 and T12 were despatched to Lyttelton, singly by various ships from London between 6 September and 23 October.

T11 (not T12 as stated in NZRO 195) was held back by the makers "as some trials are about to take place, the engineers propose to keep it for a week or two to show what can be done by it and to prove to the colonies as well as the people of England the advantages of steam power on tramways and the excellence of the engines manufactured" as the *Lyttelton Times* put it.

The engines were built to British Board of Trade requirements. Kitson proposed to obtain Board of Trade approval for their locomotives using T11 (Canterbury Tramway Co "E", later No 5) for the purpose. The trials were run before the board's Inspector, General C. S. Hutchinson, R.E. in normal service on the

Dewsbury, Batley and Birstall Tramway from 8 November to 13 December 1879. The *Lyttelton Times* of 26 February reported the trial. "...The engine commenced to run on 8 November 1879 and with a two horse car capable of carrying 40 passengers, performed an average daily duty of 62 miles to 13 December – that is 31 clear working days giving a total mileage of 1922 miles. The engine took its turn with the horse and never missed a journey....In fact for three or four days, during an almost continual fall of snow, the horse car traffic would have been entirely suspended had not the engine been available for keeping the rails clear ... Its performance at Batley has resulted in arrangements being made for the working of several tramways with this class of engine ..."

T11 was borrowed for much longer than "a week or two". In fact 14 months were to elapse before the ship *Gareloch* left London on Christmas Eve 1880 with the engine on board, with brand new steel tyres fitted and a peace offering of steel tyres for the previous four engines. Kitsons must have literally run the wheels of her as a demonstrator!

With the Board of Trade's approval gained for the engine, the time was now ripe for a really extended test. Kitson induced the Leeds Tramway Co to apply to the Board of Trade for three month's trial of a locomotive on their lines. A new engine, T15, similar to the New Zealand engines, but with slightly larger cylinders, was built for the trial which began on 14 June 1880. Now, according to Kitson records, T15 did not leave the works until 17 June and T11 was still in Leeds. It seems very likely that both engines worked the trial and being run into the works each evening, no one knew the difference. Speculation perhaps, but there is some evidence to the effect that there was more than one tram engine in steam in Leeds at the time. It was extremely important to Kitsons that the engine perform perfectly.T15 went on to make a further 12 month trial in Leeds, commencing late in 1880. The engine ran over 15,000 miles and scarcely missed a trip. This successful trial led to the adoption of steam power in Leeds and subsequently on other British tramways generally.

The Canterbury Tramways Company's system not only expanded rapidly but traffic exceeded expectations. A full four-engine timetable was worked with only four engines to hand. While there was a demand for Sunday services this was resisted – not so much as it was Lord's Day but overnight and Sunday were the only times when maintenance, boiler washouts and repairs could be undertaken. Consequently when T11, now christened E, arrived in Lyttelton in April 1881 she was immediately erected and put into traffic within days. A repeat order for three more engines had already been placed. The new engines "F", "G" and "H" (T27-29) were shipped from London in one batch in June 1881 only six months after the belated dispatch of "E".

T7, the Dunedin engine, had only a short life as a tram engine – from 15 January 1880 when she went into service until its destruction in the depot fire of 23 December the same year, James Craig, Proudfoot's engineer, regarded it as the very best of the fleet of Hughes, Baldwin and Kitson engines. The remains were sold to Conyers and Davidson of the Otago Foundry and incorporated into a vastly altered 0-4-2 saddle tank which ran for many years on bush tramways in Southland.

The Christchurch engines had much longer lives. In 1882 the full complement of eight engines A to H was renumbered one to eight. No 7 was involved in a boiler explosion in 1886 while standing unattended, but was repaired and soon back on the road. They became due for reboilering between 1888 and 1893 at a time when the Canterbury Company was in a poor financial condition. No 4 was scrapped in 1893 but the others were reboilered by the newly formed Christchurch Tramway Company. The Christchurch Tramway Board bought out the several tramway undertakings from 1903 to 1906, and from May 1905 instituted a policy of steam services to the greatest extent pending electrification. A second-hand Baldwin engine, No 9, was bought from the New South Wales Government Tramways in 1906 to assist.

Electrification of all lines was completed in September 1914. There was then only occasional special and peak passenger work for the steam engine whose normal duties were confined to track and overhead maintenance trams and quarry shunting. Accordingly, Nos 1, 2 and 5 were withdrawn but Nos 3, 6, 7 and 8 were reboilered once more in 1928-1930 to continue in this work. There was very little of it, No 3 being scrapped in 1944 while 6 and 8 followed in 1950. No 7 was retained as a relic and outlasted the Christchurch Tramway system which closed in 1954. After some vicissitudes it became the property of the Tramway Historical Society at Ferrymead Park. It has been completely overhauled and has run intermittently on the Ferrymead Tramway for many years now. It is probably in as sound mechanical condition as it ever has been and was recommissioned following her 10-year survey on Sunday 4 May 1997. There is every reason to expect No 7 to continue to work special services at Ferrymead for many years to come.

With electric tramways in general operation soon after the turn of the century one could be forgiven for thinking that there would be no further developments involving steam tramways engines. Far from it. Further engines, albeit second-hand, were acquired for tramways in Christchurch and Wanganui and two

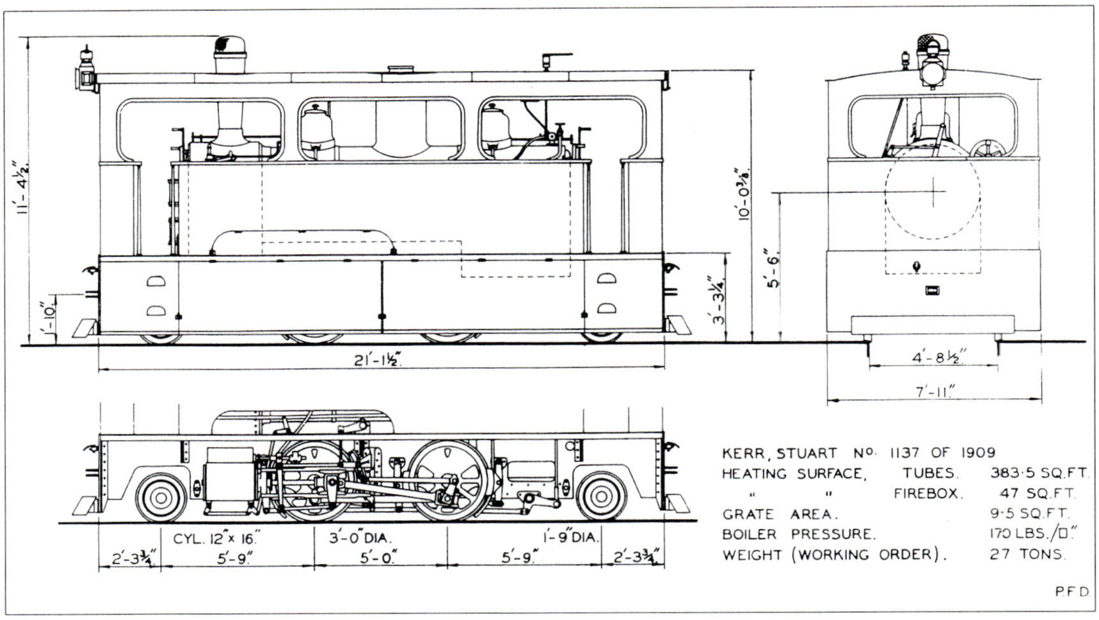

One of the last steam tramway engines built in England, Kerr Stuart No. 1137 of 1909, No. 2 of the Takapuna Tramways & Ferry Company Limited. *Drawing by Peter Dyer*

new steam tramways were inaugurated. One, the Otatara Tramway at Invercargill, was essentially part of a suburban land development using a third-hand Merryweather locomotive and cars all originally from Wellington. The other, the Takapuna Tramway and Ferry Company, serving Auckland's North Shore, was a well-built undertaking which intended steam power as an interim measure only pending eventual construction of a powerhouse and electrification. In some ways it resembled a railway with powerful locomotives, rear banking of three and four-car trains up the 1 in 25 grade from Bayswater Wharf, Westinghouse automatic air brakes and reputed speeds of up to 40 mph through the dips.

The initial Takapuna locomotives came from the British firm of Kerr Stewart and Co in 1909, a firm not noted for tramway locomotives, but the established builders, Merryweather, Kitson etc had ceased interest in tramway matters. The original locomotive No 1 (Kerr Stewart 1096) though used on construction was probably not powerful enough for full day-to-day service with its 9½ inch by 15 inch cylinders and was quickly followed by No 2 (K.S. No 1137) with 12 inch by 16 inch cylinders, and a higher boiler pressure. No 3 (K.S No 1216/1911) was even more powerful with 12 inch by 20 inch cylinders. All three were 2-4-2 side-tank engines with all-over cabs. These must have given the company satisfaction as the final engine No 6 (K.S. No 4301/1924) was similar in dimensions to No 3.

Baldwin engines were yet to play a part in Christchurch, Wanganui and Takapuna. We have already noted some early Baldwin productions in Dunedin. Like most steam-tram manufacturers Baldwin had early substituted a locomotive-type boiler for the initial vertical boiler. The Baldwin dummy was basically an industrial saddle tanker encased in an all-over cab that resembled a horse car. It was turned out in various sizes from 9 inch diameter cylinders to 12 inches diameter and even offered as a Vauclain compound. It was very powerful for its size and speedy but with a nose heavy weight distribution which combined with the nosing due to the piston thrusts could cause the engine to roll alarmingly at certain critical speeds. It was easily capable of 40 mph and more.

The type came into its own in Sydney from 1879 and became the standard New South Wales Government Tramways engine. The rough riding was noted and hoping to cure this the New South Wales Government Tramways ordered some 10-inch motors with a trailing truck – thus a 0-4-2ST in 1881. Perhaps these were not a success as the trucks were removed in 1887. One of these engines came to Christchurch in 1906. As Christchurch No 9 its Sydney identity was lost on sale. It worked Christchurch Exhibition passenger services in 1906 but as electrification proceeded it was relegated to maintenance work, being finally condemned in 1923.

Sydney had largely standardised on Baldwin's 11 inch by 16 inch motor. No 100 (Baldwin No 11665/1891) of the final Baldwin batch eventually went to the Gonville and Castlecliff Tramway Board in 1910 as a construction locomotive and was taken over by the Wanganui Corporation on the absorption of these suburban boroughs in 1925. For some months following a powerhouse breakdown, it provided the sole Wanganui tram service. Eventually it came to MOTAT, has been reboilered and completely overhauled

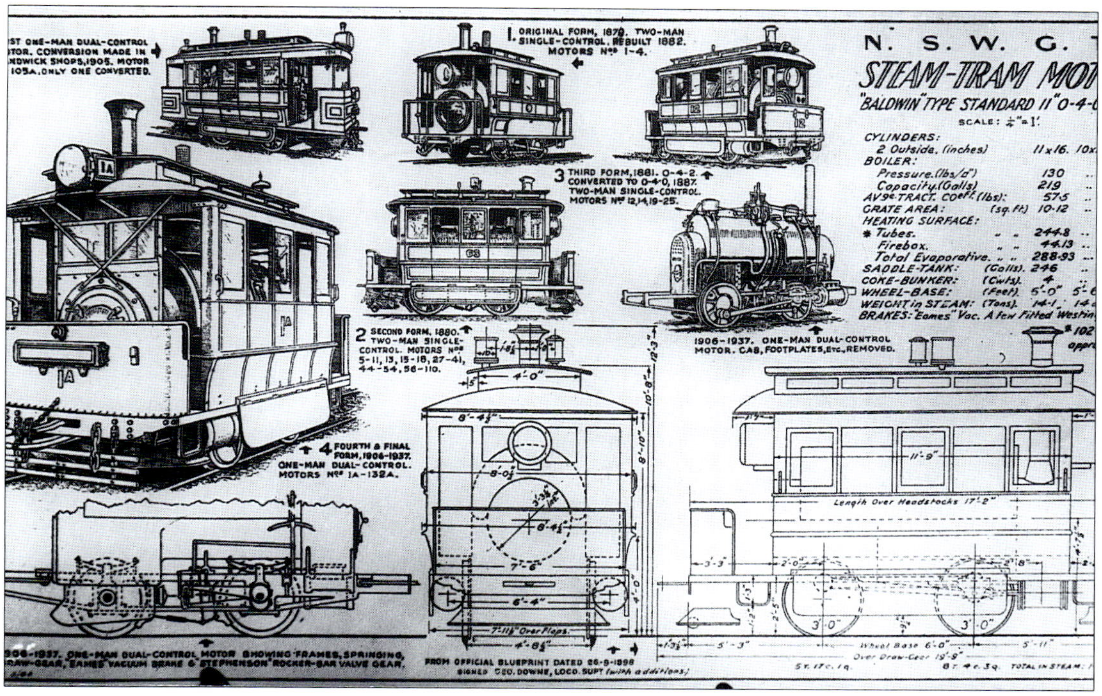

Baldwin steam-tram motors as running on the New South Wales Government Tramways. The Sydney system (NSWGT served other towns) was one of the largest steam tramway networks in the world. *Australian Railway Historical Society*

and was recently recommissioned.The Takapuna Baldwins, Nos 4, 5, and 6 with their 12 inch by 18 inch cylinders, were even larger machines than those of Sydney. With a tractive effort of 10,070 lbs at 80% boiler pressure and an all-up weight of 22½ tons they were of similar power to the large Kerr Stewart machines but probably due to better valve proportions, were reputed to be speedier. No doubt they rolled like their Sydney cousins. Probably the Kerr Stewarts with their leading and trailing trucks rode and tracked better and this is possibly the reason for the preference for the British product when the final engine No 7 was ordered.

New Zealand then has been host to the leading tramway engine manufacturers of Britain and the USA, being Merryweather, Kitson and Baldwin. The only other major British tram engine was the vertical-boilered geared Wilkinson patent engine which did not run here. If it was considered, it would have been rejected as too slow and not powerful enough. William Wilkinson, however, is reputed to have developed his engine from a Chaplin vertical-boiler engine in 1881 New Zealand's street tramways commenced in Thames in 1871 with a Chaplin locomotive.

The steam tramway engine became a practical proposition during the final quarter of the nineteenth century only. Even as the builders perfected their standard designs, Siemens had demonstrated the practicality of electric tramways. It remained only for Siemens, Edison, Sprague and others to improve its economy and thus scupper the steam motor. That did not take very long – a mere ten years or so. On some of the larger systems, such as Sydney or Paris, the steam engine continued for a time on minor routes considered as not worth electrifying. Like Takapuna they finally succumbed to the motor bus.

THE GRIPS, CABLES & POWERHOUSES OF THE DUNEDIN CABLE TRAMWAYS

Gerald Ditchfield

Gerald Ditchfield is a retired Mediator member of the Employment Tribunal, a former Industrial Conciliator. From riding the Stuart Street cable line as a pupil of the Otago Boys' High School, he has made a study of the cable tramways of Dunedin, photographing, interviewing employees and recording in detail all aspects of this unique part of New Zealand history. He was President of the New Zealand Railway & Locomotive Society for 1972-74. When the Roslyn cable line closed in 1951, he had the honour of turning off the steam winding engine – the last steam powered street cable tramway in the world.

THE GRIPS

At the very heart of all public passenger cable tramway operation lies "The Grip"; it is the link between the moving traction cable and the otherwise immobile cable tramcar. In basic terms it is a device fitted to the cable tram which, by means of an arm or shank projecting downward through a slot in the roadway, connects the tram to an endless wire-rope which is hauled through a conduit below the surface of the roadway. To the lower end of this arm is fitted a cable gripping device that enables the driver, or "gripman", to control the movement of the tram. By backward (or forward)* movement of a lever, the grip takes a gradual, vice-like hold of the moving cable to start the tram. To halt the grip tram the traction cable is released by appropriate* movement of the same lever; this then permits the brakes to be applied to stop the tram at any point. When "in release" the cable is permitted to run unimpeded through the grip.

Andrew Smith Hallidie is generally credited with developing the first practical street cable tramway grip; this occurred on the Clay Street Hill Railroad, San Francisco, in August 1873. Actuated by a pair of horizontal wheels fitted to a vertical screw-threaded shaft, one of the wheels raised or lowered the height grip while the other by means of an internal spiral winding, controlled the opening and closing of the gripper jaws. Wheel-operated grips, however, were relatively slow acting and from around 1877 were displaced by the more quickly responding, lever-actuated grips. Both these styles of grips were to see service during the 76 years that cable tramways were in operation in Dunedin.

However, before considering the various grips that were adopted in Dunedin some important points need to be noted. The grippers used by the Dunedin cable trams were always located at the centre of the tram or "dummy" in a gripman's compartment or "box" from which passengers were excluded. Although required to travel in both directions – either "up hill" or "down hill" – Dunedin's 4-wheeled cable trams were in effect "single enders". To reverse direction at each terminus, the gripman merely moved around the tram's various centrally located controlling levers in order to face the required direction of travel.

It should also be noted that within a Dunedin context, reference to a "top" or "bottom" grip refers to which particular gripper "jaw" was mobile during the action of opening or closing the grip. e.g.: A "top" grip was one where the gripper's upper jaw(s) was mobile and gradually took hold of the moving cable by bearing down firmly on the grip's fixed, lower jaw; reverse action took place when releasing the rope.

Roslyn Grip

When the Roslyn Tramway Company opened on 6 February 1881 it was the first cable tramway to be constructed outside the United States, although some four installations were already operating in San Francisco. Until recently it was thought that the original grips used on the Roslyn line were the same or similar to those used by Hallidie in 1873 on his Clay Street line. However, some recently located original drawings by George Duncan, Engineer to the Roslyn Tramway Company, reveal a grip of quite radical design and operation. While Hallidie's basic wheel and screw style of operation applied, Mr Duncan by incorporating a system of horizontal spur gearing, was also able to open and close the jaws of his grip while simultaneously applying or releasing the track (or slipper) brakes through single movement of a horizontal

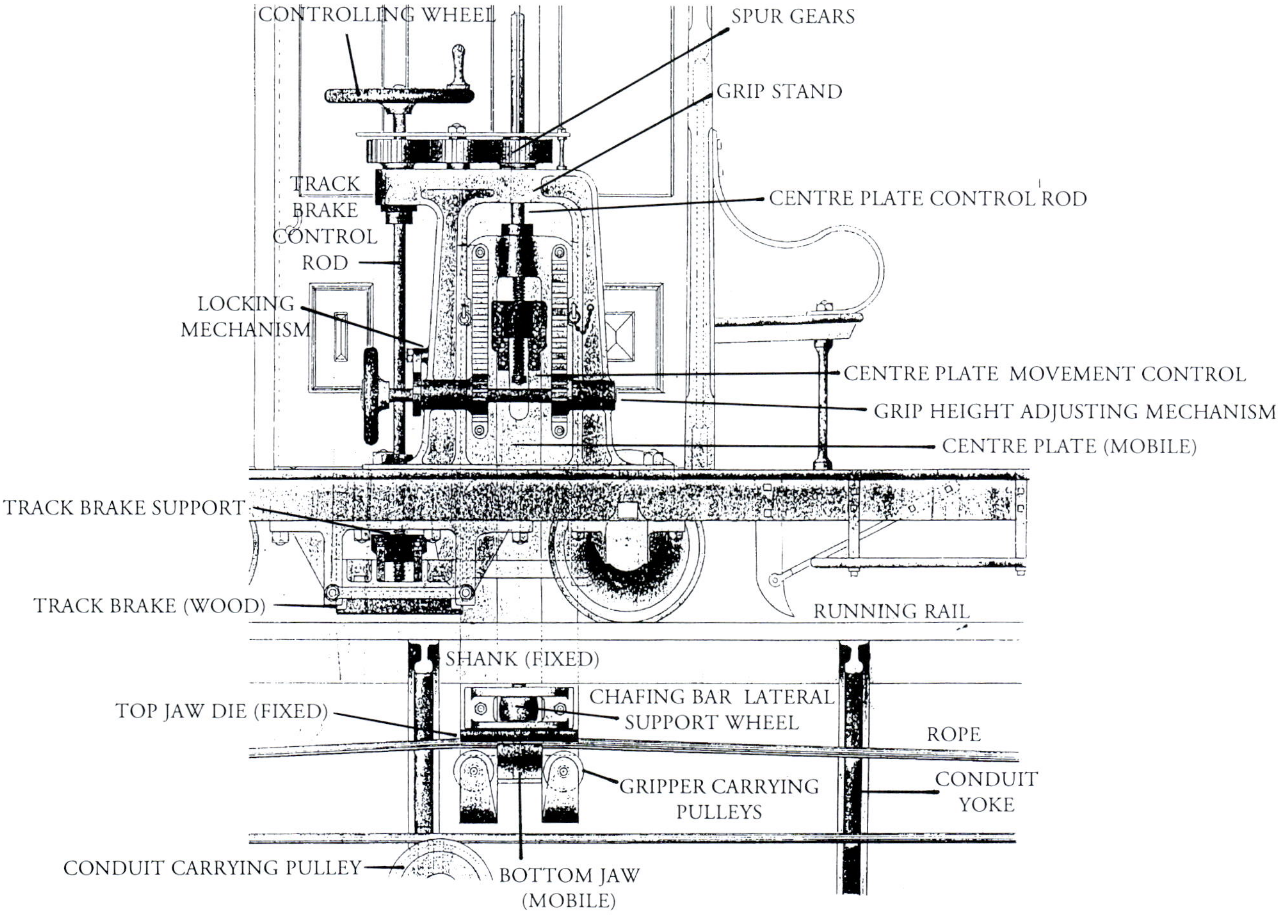

A plan of the combined grip and track brake as designed by the Dunedin engineer George Smith Duncan for the first cable street tramway in the southern hemisphere. Believed to have be used on the grip cable-trams on the pioneer Roslyn Tramway Company line which opened in 1881.

wheel. The height of the grip within the conduit was adjusted by a horizontal screw, mounted on the base of the grip stand. This grip appears to have been a double jaw, "bottom", side grip.

While it cannot be stated with absolute certainty that such a radical design was given practical effect on the Rattray Street line, it is more than probable that given Duncan's outstanding talents, including his invention of the "pull" curve, that the gripper illustrated was in fact used. Further, a detailed report of an accident which occurred on 30 April 1881 confirms that at least one of the cable trams made use of a screw-operated track brake. The evidence available suggests that these wheel-actuated, combined braking and double jaw, side grips were in use until cl906.

A change to lever-operated grips was foreshadowed in 1902 when cable grip tram No. 6 (first series) was experimentally fitted with a lever-actuated side grip of a type which was reported as having been developed in Melbourne. This grip was fitted with a lever-mounted "set" adjusting, screw thread which enabled the gripman to adjust "the set" of the gripper's double jaws while in traffic according to the number of passengers being carried and the amount of wear on the cable. This innovation was successful and became a standard fitting for the grips installed on the second and final series of Roslyn cable grip trams constructed during 1906 1908.

Mornington "Mainline" Grip

Although only two years separated the opening of the Rattray Street and High Street cable tramways, technological advance was evident in the type of grip adopted by The Mornington Tramway Company Ltd. The one mile route which was also engineered by George Duncan, was ideally suited to cable traction, being practically a "straight pull" up High Street with only a double-tracked, 250' radius "pull curve" required into Eglinton Road. The knowledge and experience gained by Mr Duncan in 1882 from his personal contact with other cable tramway engineers in North America saw adoption of contemporary American cable tramway practice for the Mornington line, including right hand street operation.

Photographic evidence shows a gripping mechanism that was lever operated and featured a large quadrant which enabled the gripman to lock or hold the gripper jaw in its various operational open or closed positions. Described as a single jaw (bottom) side grip it is believed to be the same style of grip that had been developed earlier in San Francisco by Henry Root and became known as a "California" grip.

Initially, this grip had drawbacks. When approaching either terminus at the foot of High Street or the Mornington township, it was necessary for the cable to be physically extracted from the gripper jaws by means of a "rope hook". At the foot of High Street this was relatively easy, the track being on a down grade so that the tram was only required to move across to the "up" line by gravity. At the Mornington terminus this was not possible and required the crew to push the tram up a slight rising gradient onto a turntable, correctly line the car up with the turntable off track and then push the tram off the turntable onto the "down" line, before physically placing the rope back in the jaws of the grip once more; these tasks were both time consuming and laborious.

In 1888, the recently appointed (1887) Tramway Engineer, Joseph Lowden, effected considerable improvements. The grip was modified so the rope was automatically ejected from the grip when approaching either terminus; points and crossovers were installed at Mornington while a track "dip or depression" was also constructed on the "down" line so that the rope was automatically fed into the open jaw of the grip. At the Exchange terminus a new vertical sheave was also installed so that here the cable was fed into the open jaws of the upbound car's grip, at the correct height.

Another disadvantage was that the lever used to operate the grip when the tram was travelling down-hill, had to be pulled back against the downward inclination of the tram in order to take hold of the cable and pushed forwards against the car's inclination when travelling up-hill. It is understood that the practice continued until 1911 when the size of the gripper quadrant and its ratchet teeth were reduced, and operation of the lever(s) was reversed in order to assist the gripmen. Other minor modifications may have also been made when the speed of the cable was increased from 7.95mph to 10.66mph in 1925. In 1928 when the "mainline" changed from right-hand to left-hand street operation it is likely that some other operational changes may have been required; however, such details are not presently known.

Maryhill Extension Grip

Because of engineering and operational considerations, it is believed that a modified "California" type grip was introduced on the Maryhill Extension. Built as a single line with a mid-point passing loop, configuration of the cable required the traction rope to be operated as if it were two separate ropes travelling in opposite directions within the same conduit; this produced its own engineering problems. A double jaw (bottom), side grip was chosen which enabled the single tram used (only one tram was ever in traffic at a time) to operate in both directions using as appropriate, either the "in-bound" or "out-bound" rope. The grip was also fitted with rope carrying pulleys at either end of each set of "bottom" jaws; this was to minimise the wearing effect of the cable on the jaws soft metal "dies".

At the Mornington township terminus, the "out-bound" rope was manually placed in the gripper by means of a "rope hook" . The gripman then signalled the engine-driver to start the cable (which was stopped during off-peak periods) and having satisfied himself by "sound and vibration" that the rope was correctly "in the jaws", would start the car by taking a "full" grip of the rope to descend the 1 in 3.5 reserved track gradient to Glen (now Glenpark) Road. On approaching the passing loop where the "two" ropes separated, the gripman would let go the "outbound" rope and coast into the loop which was a straight run in, curve out variety; if necessary the tram would halt to set down or pick up passengers. The gripman then took "partial" hold of the "out-bound" cable in order to negotiate the remainder of the loop before again taking a "full" grip in order to enable the tram to continue its journey to Maryhill terminus. On approaching the terminus at Mitchell Avenue, the gripman would "throw" the cable which was automatically ejected in order to permit its reversal by the "terminal sheave" located beneath the roadway. The tram came to a halt at a specified rope pick-up point where the now "in-bound" rope was automatically fed into the grip for the car's return to the Mornington township, during which the procedure was repeated in reverse. On return to the Mornington terminus the gripman on surmounting the gradient, would immediately release the "inbound" rope to halt the tram; the "in-bound" rope was then directed back into the powerhouse.

Elgin Road Extension Grip

Although the Mornington Municipal Tramways extension along Elgin Road to Balaclava was ill fated, it closely matched its counterpart to Maryhill. Essentially, both routes served the same area, and were roughly parallel to one another; each was constructed for single track, passing loop operation; three such passing loops were required on the new route. Moreover, the route chosen to Balaclava was a very tortuous one involving very tight curvature that required no less than six "pull curves", two of which (Havelock Street into Elgin Road) were effectively, right angled. Another (Havelock Street into Mailer Street) was required to ascend a short 1 in 7.9 gradient when returning to the Mornington Township via Lawrence Street and became in effect, a "reverse pull curve". Overall, these kinds of operational circumstances plus the Borough Council's alleged purchase of "cheaper" ropes, were said to have contributed to the short-lived nature of this line's operation.

However the double jaw (bottom), side grip used on the Maryhill route was also suited to operations on the Elgin Road route and it was this type of grip that was fitted to cable grip tram No. 6 when this was specially constructed in 1906 to operate this route. Few other details have emerged, however, regarding the operational aspects of the grip used on this route.

Kaikorai Grip

By the time the Octagon to Kaikorai cable tramway opened in December 1900, the Dunedin & Kaikorai Tram Company Ltd, was "spared the agony of experimentation" of its predecessors. Moreover the route constructed was only moderately curved and contained a series of stepped gradients, the maximum of which near Highgate was 1 in 4; this was later eased to about 1 in 7. The type of cable tramway rolling stock introduced and the operational practices adopted drew heavily on the experience of their Dunedin counterparts. It is believed that all curves on this line were of the "pull" variety.

Lever-actuated grips were installed from the beginning but differed from those used by the other Dunedin lines; although double-jaw side grips were adopted, only the top jaw(s) were mobile. Designed and patented by Mr. J Roberts who was responsible for engineering this line, the grip was reputed to be difficult to work and often required the combined efforts of the gripman and conductor (pushing his foot hard against the lever) to secure a proper hold of the rope when an ascending, heavily laden, grip car en route from City to Kaikorai Valley during the evening "peak", was in traffic.

A further disadvantage arose from the Company's failure to provide a "gripper trough" inside the tramshed. This obliged all tram crews to raise or lower the heavy grip by means of a portable, hand-operated winch, above/below the level of the carshed floor whenever a grip tram was required to enter or depart the depot. Although only intended as a temporary measure, the practice was to remain in force right up to closure of the service in 1947; it was a much despised task.

THE CABLE

By the time the cable tram emerged in 1873, wire rope was already in use for haulage and other purposes in a wide range of industries that included railways, coal mining and other activities. Indeed the cable tram's inventor, Scottish-born Andrew Smith Hallidie, had been involved in the business of wire rope manufacture in San Francisco since 1857. Street tramway engineers could thus be offered from the catalogues of British or American manufacturers, similar sets of alternatives even if the cables available differed in terms of diameter, weight, number of strands and the directional lay of their windings. The steel cable was commonly known as "the rope".

The builders of Dunedin's pioneer cable tramway (Roslyn–1881) Reid & Duncan who since 1879 had possessed ". . sole rights to construction of wire rope tramways in the colony of New Zealand. . . " no doubt drew on the experience of earlier established San Francisco counterparts when considering their choice of cable. Similar circumstances existed for the other Dunedin installations that were to follow, but local experience became an important factor. Over the years the various Dunedin cable tramways operators came to adopt standards and specifications that were the same, or similar to each other in their choice of traction cables.

Records that exist reveal that overwhelmingly, the Dunedin cable tramways operators chose British manufacturers as their suppliers of wire rope; the names of George Craddock, Alan Whyte, Whitecross and T. & W. Smith appearing quite frequently. It is believed that during World War II when steel ropes became very difficult to obtain, some cables had to be purchased from North American sources. Also during World War I the Roslyn cable line found it difficult to obtain new rope from British sources. The ropeman, Archie "King George" McLeod, is said to have made up a replacement rope for the Rattray Street route by using pieces of old ropes that were "lying around". He received a bonus of 5 pounds for this work!

Rope Specifications

Being a close-knit community and in latter years with only a single management, (the Dunedin City Corporation Tramways), specifications set down for the endless wire ropes that drove the Mornington,

Maryhill and probably the Kaikorai cable tramway systems, were effectively standardised. On these routes each rope consisted of a compound of six "strands", each "strand" containing 13 wires (i.e. 7 wires over 5, over one small flexible wire) using a "Lang lay" which involved winding the wires in each strand in the same direction as the winding about the "rope's" central, hemp core; each strand ran full length of the rope. The "Lang lay" also enabled the cable to wear smooth as it aged, thus presenting a long continuous surface to the grip over which pressure was evenly distributed. "Tru-lay" ropes were introduced in the mid-1930s; although more expensive, this type of rope was said to last longer. All ropes had a circumference of 3.5"; a tensile strength of 67.5 tons psi and a guaranteed breaking strain of 29 tons.

Exceptions existed on the Rattray Street (Roslyn) and Elgin Road lines where their respective rope circumferences were recorded as being 3.625" and 3" and they respectively weighed 13.921b and, 91b per fathom. On the Elgin Road line the rope consisted of 6 strands of 15 wires (i.e. 8 outer wires laid over 7 smaller inner wires); Lang's patent lay was used; breaking strain of this rope was 28 tons. No "make-up" specifications or breaking strain figures for the Roslyn line have been located.

Life of a Rope

Once installed, estimating the life expectancy of any cable tramway rope was very difficult to predict; however, some "ropemen" and powerhouse engineers became quite skilled at this by observation and their "listening to the song of the rope". Factors that governed the life of a rope included frequency of service; sharpness of curves and steepness of gradients; speed at which the rope operated ; loadings to which it was subjected to and, in the case of Dunedin, the severity of that city's winter weather. Quality of the steel used to manufacture the cable was also important as were operational incidents caused by inattention of the gripmen or other factors. For Gripmen "cutting the cable" was the worst possible crime of all; dismissal or demotion usually followed.

Constant surveillance of the rope had to be maintained at all times; broken wires or loose cable stands that could get caught in the grip or cause "bunching", were frequent reasons for repair. As soon as detected, these faults had to be corrected and repaired as quickly as possible to avoid further damage to the rope. Such repairs could be required suddenly, with the work, regardless of weather conditions, taking place, if possible (at night or weekends), after cessation of time-tabled services. In the case of the Roslyn line some of this work might be carried out under cover within the confines of its generously proportioned Powerhouse. However, on the Mornington, Maryhill lines and presumably Elgin Road, such work had to be performed outside and occasionally in frosty weather or during winter snow storms. In these circumstances the hand tools used, sometimes had to be warmed up before use, by being placed before the Mornington depot staff lunch room open fire.

In the main, the Dunedin installations were able to achieve good service from their ropes. Records show that in 1894 the Mornington Tramway Company (opened 1883) achieved 3 years 4 months and 25 days and 172,391 miles running from the second rope that it installed on the "mainline" and 3 years, 7 months, 5 days and 180,234 miles for the rope taken out of service on 13 July 1893. On the Maryhill Extension (opened 1885) the same record shows 2 years, 8 months, 8 days service and 33,550 miles. Also the rope taken out on 8 February 1894, operated for 6 years, 2 months, 10 days and 53,752 miles. Both these ropes were supplied by George Craddock and Company of England.

On the Rattray Street (Roslyn) line up to 1894 their 7000' rope, also supplied by Craddock & Co. ran for periods that lasted from 678 to 726 days; however, the ropes (2) (supplier unknown) which preceded and succeeded those cables lasted only 292 and 462 days respectively. On the Kaikorai (Stuart Street) line their best cable (12,605') which weighed 12.5 tons and was also supplied by Craddock & Co, is recorded as having lasted for 671 days, travelled 139,660 miles and conveyed 2,123,980 passengers. Like the Maryhill Extension the Kaikorai cable line ceased the between-trip operation of its cable during off -peak periods.

The shortest cable life expectancy was on the Mornington-located, twisting Elgin Road line, where the rope's life expectancy was exceedingly short. The first 6,550' rope installed (19 September 1906) lasted until 25 July 1908 in spite of having to wear the "rough spots off" a brand new installation. This rope ran 11, 859 miles. As no spare cable was on hand, the service was suspended until 10 November 1908 when another new rope was installed; after travelling some 14,448 miles this rope was replaced on 3 September 1909. That rope – the final to be installed – lasted only four months before it too became unfit for further service; all cable-hauled timetable services on this line ceased on 22 January 1910.

Rope Installation and Replacement

Typically, new cable tramway ropes were both expensive to buy and heavy to handle. In Dunedin it was standard practice for a new rope to arrive there on a wooden shipping reel that was then hauled up the steep streets of the city to the Powerhouse concerned. Teams of up to 20 horses (later traction engines and motor lorries) were used for this purpose.

When laying the very first cable, this could prove quite difficult; either of two methods of installation as practised overseas, may have been adopted in Dunedin. One involved the harnessing of horses to a grip tram which would hold one end of the new rope tightly in its grip- the tram was then hauled through the system until both ends of the new rope met, where its formation was then made endless, by being spliced together. More than one grip tram might have to be used. Another similar method might have been adopted, except in this case, the two ends of the cable were hauled separately (again by teams of horses) from the line's highest point (usually the Powerhouse), down each track for splicing which would then take place at the line's lower terminus. It is not clear exactly what method was adopted for the original Roslyn or Mornington cable lines.

However, in the case of the Rattray Street route, it is known that it took two men crawling through the conduit on their hands and knees, assisted by teams of horses, two days to thread the 7000' cable over and around that line's many pulleys and sheaves. The use of horses is also mentioned in connection with laying the original cable for the Maryhill Extension, while a traction engine hauling a grip tram (#6) with the rope held in the jaws of its grip, appears to have been used to install the first Elgin Road cable in 1906. Which method was used to lay the first Stuart Street (Kaikorai) cable is unknown.

By contrast, installation of replacement ropes was a comparatively easy task. The old rope could be cut off at the Powerhouse while the new rope, using a special clamp fixed to the other end of the old cable, could be run out slowly over the line using the Powerhouse winding gear, while the old rope was reeled in by a winch onto a spare reel for subsequent disposal. Once both ends met, the new rope was then spliced into an endless formation. Constant oversight by the trackmen to correct positioning of the cable on its carrying pulleys and sheaves was essential during this operation.

Keeping the Rope Tight

It was of the utmost importance that correct and constant tension was maintained on the traction cable at all times; this was an operational imperative for cable tramways everywhere. Variations in tension would arise from in-service cable trams, up and down the line, constantly gripping or releasing the traction cable at random; changes in temperature could also vary the length of the rope; stretching and narrowing of the cable due to operational wear was another factor. Stretching of a newly installed cable could be substantial. On the Dunedin lines this might be up to 12'-20' after only one day's running of a brand new cable; further stretching would occur during the following week and beyond.

Types of Tension Adjustment

Within the Dunedin environment two basic types of automatic tension adjustment were applied. One consisted of a terminal sheave mounted horizontally on a small carriage that was free to move backwards or forwards a short distance on rails, subject to the tension applied by means of a counterweight that hung over a pit. The rails referred to were set on an incline and the entire apparatus was housed at the line's lower terminus in a conduit "vault" below the road. Such a "tension pit" was installed on the original Rattray Street (Roslyn) when it opened in 1881. The "tension pit" was located at the junction of Rattray and MacLaggan Streets; an 11' terminal sheave and 15' deep counterweight pit were used at this location. The site was "rediscovered when the Rattray Street line was being finally ripped up in 1953.

The Mornington Tramway Company also used this kind of arrangement when it installed "tension gear" in the vaults located at the terminals of its High Street and Maryhill extension lines. The Elgin Road line (built by Mornington Municipal Tramways), however, appears to have been fitted with only a single, fixed sheave; presumably the necessary tension adjustments were somehow taken care of within that line's "rope -race" located inside the Mornington Powerhouse.

The other type of tension gear used consisted of a vertically mounted "Idler Sheave" fitted to a four-wheeled "tension carriage" that moved on rails astride an internally located powerhouse "rope race". As the cable slackened or tightened, the moveable carriage which was governed by a counterweight, was free to move backward or forward automatically.

This method was adopted by the Roslyn Tramway Company when it extended its route to Kaikorai Valley in 1900; however, this had to be replaced in 1911 when the company installed a 325hp Tangye steam winding engine to cope with traffic generated by its other subsequent extension into the same area. Some changes were made to the tension arrangements. Two cable driving wheels were purchased second hand from Melbourne cable tramways; the cable entered the powerhouse and before travelling on to the cable-driving mechanism, was reversed over an adjustable, counterweighted (5.5 ton) tension carriage astride the

80' powerhouse "rope race". From here the rope made a "double wrap" (two turns) around a large fixed, grooved, wooden segmented driving wheel that was fed by a "floating pulley" which weaved back and forth, in order to maintain correct tension on the cable and to eliminate any slippage on the part of the rope as it passed through the winding gear. When first installed this "new" method of winding the cable became cause of a 50 pound "bet" that the driving system would collapse after two months. The arrangement proved most durable and was still in use when the Roslyn line closed in 1951.

Another similar, but different, type of tension carriage arrangement was installed on the Stuart Street (Kaikorai) cable line in 1900; this installation is believed to have been a "double tension carriage". In this situation a tension carriage complete with its own tension ("idler") sheave and running rails, was placed on top of a larger and heavier under-carriage that also moved on its own set of rails.

This under-carriage was not free to move back and forth on its rails automatically; it could be shifted by the Powerhouse staff and its stopping point was fixed by a rack and pawl arrangement. Stretching of the cable was compensated for by moving the heavier, lower carriage, while the smaller carriage with its own tension gear positioned on top, was still free to move back and forth on its own rails to take care of the other, but smaller, variations in tension.

This form of tension adjustment was well suited to situations where space within the powerhouse was a consideration. Also any gross lengthening of a new rope that required backward movement of the lower carriage, could be accommodated. In this circumstance an adjustment of say only 25 feet, could then be made account for as much as 50 feet of "slack" in the rope.

This "double carriage" type of tension arrangement is used by the cable tramways operating in San Francisco.

THE POWERHOUSES

In the operation of any endless wire rope, passenger street tramway next in primacy to the Grip and the Cable, was the Powerhouse. Here was housed the machinery and other devices that both propelled the cable through its underground conduit and in many instances kept the traction rope under correct tension at all times. The Powerhouse was also a place where the wire rope traction cable could be kept under constant surveillance for any sign of fault, "stranding" and general wear.

For haulage purposes cable tramway engineers preferred a direct route with as few curves as possible. Fewer curves reduced the friction on the rope which in turn reduced the horsepower required; around 60% of the horsepower required could be absorbed by just moving the rope without any trams attached. The amount of wear on the moving cable was also reduced. Topographically, Dunedin was most fortunate in this regard, for its cable routes (Roslyn, Mornington, Maryhill and Kaikorai) were essentially direct pulls up to these hilltop suburbs; generally only moderate curvature was involved.

The exception to this was the Mornington-located Elgin Road Extension which although essentially level, required the rope to be hauled through six major curves, two of which were at right angles (Havelock St into Elgin Road and return). Another included a track configuration in Lawrence Street on a 1 in 7.9 gradient which was in effect a reverse "pull curve". This line it is believed, was among the most tortuous in the world, notwithstanding the size and complexities of some other overseas installations.

Technology had already developed the stationary steam engine to a high level of perfection and reliability. The constantly varying loadings placed on the winding machinery caused by the grip cars taking hold or letting go of the traction cable at random, required motive power that could cope with such demands while maintaining a constant rope speed. The low revolution, stationary steam engine (compound or simple expansion) usually met those criteria. With two exceptions, all of the cable tramway installations built in Dunedin made use of steam throughout their operating lives. These exceptions occurred on the Maryhill Extension and Mornington (High Street) lines, where electric power replaced steam in 1921 and 1925 respectively. The Roslyn line was the last steam-powered ,endless wire rope tramway in the world when it ceased operation in October 1951.

Dunedin & Roslyn Tramway Company

Original Powerhouse: 1881- cl90l: Powered the original cable line which ran from the corner of MacLaggan & Rattray Streets to Town Belt terminus.

Location of Powerhouse: End-on to Town Belt terminus.

Motive Power: 16hp, Marshall portable engine fitted with Hartnell's patent automatic throttle, capable of developing 40hp. A belt from engine flywheel drove winding gear which drove traction rope at 6mph.

Comments: The Powerhouse was serviced by two (non-passenger) work cars; a water grip tram which operated until high pressure water supply to this area was established; a coal grip tram that also hauled coal supplies from the town to the powerhouse.

Second Powerhouse: Drove second extension cl90l-l906 that operated from Town Belt terminus to Kaikorai Valley via Ross Street, Highgate, Belgrave Crescent and James (now Falcon) Street; also extension from junction of Rattray & MacLaggan Streets to Princes Street intersection. From 1906-1911 also housed cable-winding machinery for company's second (but final) extension into Kaikorai Valley but increased traffic generated by this new route soon resulted in overloading of the Marshall engine (see below). Slippage of the traction rope around the winding gear would also occur.

Location of Powerhouse: Castle Street (now Frasers Road) Kaikorai Valley, at right angles to terminus and adjacent to Car barn.

Motive Power: One drop valve 150hp, Marshall compound steam engine driving spur-geared cable-winding machinery. Also one drop valve, 150hp Marshall compound installed alongside cable-winding engine, to rope-driven countershaft which powered belt-driven 75kw electric dynamo that for some years supplied 500 volts (DC) to company's Highgate-located, Roslyn-Maori Hill Electric Tramway.

Additional Motive Power: Extensions subsequently made to above Powerhouse to accommodate a more powerful cable-winding engine. A 325hp cross compound Tangye fitted with Corliss valve gear was installed by John Chambers & Son in 1911; high pressure cylinder 15" (180 hp); low pressure cylinder 26" (145 hp); Stroke 32"; 14 cotton (later manilla) ropes drove cable winding machinery through 28' diameter, grooved flywheel weighing 32 tons which moved cable at 7.95mph. Boilers (Joseph Sparrow) (2) of 51 hp (working pressure 140lbs psi; reduced to 120lbs psi 1938 & service reduced), manually fired, rated to burn 5.6 tons of lignite coal per day.

Mornington Tramway Company

High Street "Mainline": 1883.

Location of Powerhouse: At right angles to Mornington terminus; below level of Eglinton Road, opposite Parkhill Avenue and to left of upper Glen Avenue (unformed) and Henderson Street. This Powerhouse later housed the other engines and winding gear for the Maryhill and Elgin Road extension lines whose cables also operated from this location. The car barn (entry from Eglinton Road) was located on upper level of powerhouse structure.

Motive Power: Two 79hp Marshall high-pressure steam engines, (one held as spare). Cylinders 13" stroke 26". Driven by spur gears directly connected to clip drum type, cable-winding machinery installed by Kincaid & McQueen of Dunedin. Boilers (2): Marshall, multi-tubular; working pressure 76lbs psi; manually-fired.

Comments: Powerhouse boilers serviced by a Coal Trailer which carried 1.5 tons of bagged coal hauled by regular in-service grip tram. Water supply pumped from a natural spring near Alva Street to Powerhouse holding tanks by means of a cable-driven, pulley-operated pump which was housed in conduit "vault" near intersection of High and Alva Streets.

Maryhill Extension: 1885

Motive Power: Additional engine of same make and dimensions installed alongside Mornington "main-line" engines.

Additional Motive Power: A 22hp Crossley gas engine was installed cl9Ol to supplement existing steam engines. The gas engine could be connected to either the "main-line" or the "extension" steam engines through a countershaft and clutch arrangement. Passenger capacity of mainline grip trams was increased during 1900. During wet weather conditions the existing steam engine was sometimes unable to haul Mornington-bound, peak hour grip trams and trailers up the steeper gradients. Traffic would be brought to a halt, forcing passengers to walk.

The gas engine, steam engines and winding machinery were later damaged during Depot/Powerhouse fire on 9 February 1903; damage was reported to be "relatively minor". Engines and machinery "patched up" and "main-line" haulage resumed on High Street line 25 April 1903 with "extension" traffic recommencing around end of July.

Mornington Municipal Tramways: Elgin Road Extension

Motive Power: One second-hand 120hp steam engine (make and dimensions unknown) installed for "main-line" haulage c 1906; existing "main-line" (Marshall) engine shifted over for use as Elgin Road haulage. No subsequent mention of gas engine. Elgin Road line which opened on 6 October 1906, finally ceased operations 27 January 1910.

Dunedin City Corporation Tramways

Operation of Mornington "mainline" and Maryhill extension taken over by the DCCT on I January 1916. Maryhill extension closed following PWD suggestion (due to this line's unsafe condition) late 1916; war-time shortages of materials were cited. Fierce public debate raged for several years over reinstatement of Maryhill and Elgin Road routes. In 1917 a Commission of Inquiry found in favour of "borrowing" rails from Elgin Road line to re-open Maryhill extension; debate among Dunedin City Councillors continued. Decision to repair Maryhill extension and obtain the necessary equipment made early 1919; line re-opened for traffic 27 August 1921.

Maryhill Extension

New Motive Power: When Extension re-opened an electric motor (make unknown) of 57.5hp, 725rpm had been installed; this was adapted to drive the former "main-line" (1883) spur-geared, clip drum type, cable-winding machinery; the rope speed remained at 7.95mph.

Mornington "Main-line":

New Motive Power: 200hp Crompton, auto synchronous, electric drive was installed on 20 September 1925. Traction cable made 2 turns around 12' diameter , wooden segmented, single driving wheel; rope speed increased to 10.66mph.

Operational Comment: High Street "mainline" converted from right-hand to left-hand street operation 16 August 1928. Mornington "main-line" ceased all operations 2 March 1957.

Dunedin & Kaikorai Tram Company:

Location of Powerhouse: Alongside the Car barn, close to Taieri Road – Nairn Street terminus in Kaikorai Valley.

Motive Power: Single cylinder 175hp Tangye steam engine; cylinders 16" x 32"; 75rpm; rope speed about 7.95mph. An additional new single cylinder was installed alongside original cylinder some years later. The original cylinder was retained but was disconnected from crankshaft. Boilers (2) installed by Johnson & Sons, Invercargill; working pressure 100lbs psi; manually-fired.

Dunedin City Corporation Tramways

Taken over by Dunedin City Corporation Tramways 23 May 1939; was the last private tramway to operate in Dunedin.

Comment: Automatic stoker installed for powerhouse boiler during DCCT ownership. Plant ceased operations without a Public Works Department revocation order on 31 July 1947. Traction rope removed October 1947.

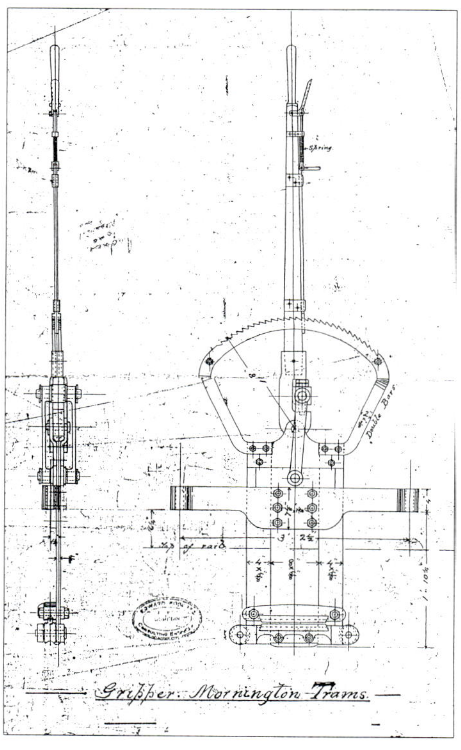

THE "PULL" CURVE & THE "DOLPHIN" BRAKE

Gerald Ditchfield

Of the many technological advances that contributed to the development and expansion of cable tramways as a form of public passenger transportation, none was to have greater impact than invention of the "pull" curve. This singular event took place in Dunedin, New Zealand, in 1881 where Dunedin-born George Smith Duncan overcame the problem of having a cable grip tram negotiate a curve while still retaining its grip on the moving cable. This breakthrough occurred on the steeply graded "Cathedral Curve" of The Roslyn Tramway Company's Rattray Street line. It was the first curve of this kind in the world.

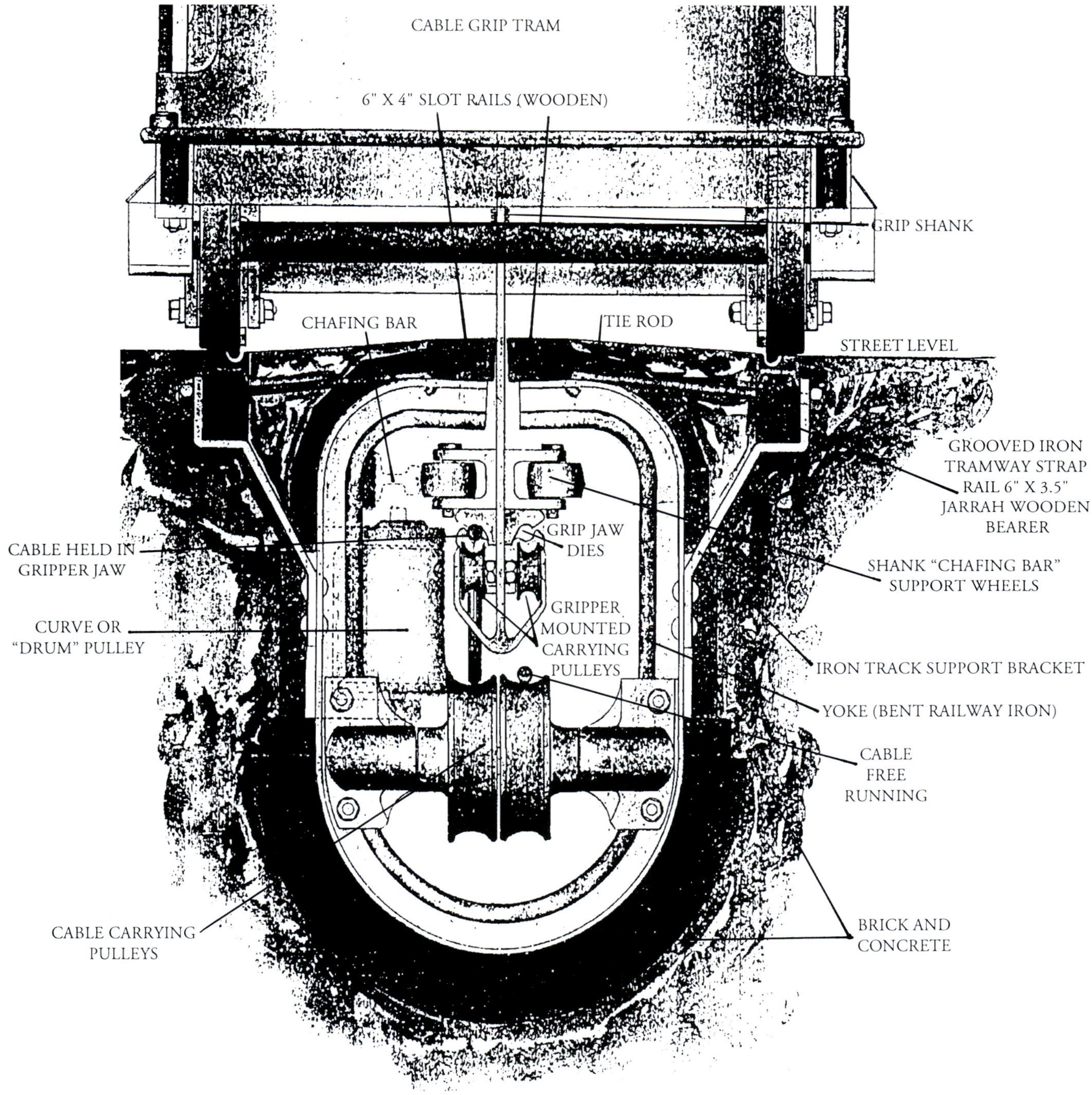

Above: A cross-section plan – left-hand view looking uphill – of the World's first cable tramway "pull" curve in Rattray Street, Dunedin, by St Joseph's Cathedral. Constructed by the Roslyn Tramway Company in 1880.

During the decade which followed Hallidie's successful establishment of his Clay Street Hill cable line in 1873, other San Francisco-based cable tramway developers and engineers considered it was impractical for a cable grip tram to negotiate a curve whilst still holding the rope. For that reason it became practice during this period to construct straight routes by following streets laid out in the usual American grid pattern. On one route, however, (Geary Street, San Francisco) this was not possible; here the method adopted was to guide the rope around the curve using a large single sheave that required the gripman to fully release the cable at the curve's commencement, to negotiate the curve under the car's own momentum, before once more taking hold of the traction cable at the opposite end of the curve. This was known as a "let-go" curve.

However, on Dunedin's original Rattray Street line which opened in 1881, this was not possible as the curve opposite St Joseph's Cathedral was located on a 1 in 7.5 gradient. It is recorded* that during 1880 Mr Duncan, engineer to The Roslyn Tramway Co, became pre-occupied "... with the problem of haulage around curves ..." for nowhere had such a difficult undertaking been attempted. He finally solved the problem by dispensing with use of the large sheave and substituting this with what he described as "drum" pulleys located behind a "chafing" bar. The cable was hauled through the curve on these pulleys while the "chafing" bar held the gripper shank away from the pulleys in order to protect it against lateral or sideways pressure. It was therefore necessary for up-bound trams to traverse the 130' radius "pull" curve at full speed for were the rope not held tightly in "full" grip , lateral pressure might tear it from the jaws of the grip, causing damage.

Such a mishap did occur on 30 April 1881 when an up-bound cable grip tram apparently had the rope torn from its grip while rounding Cathedral Curve. The resultant runaway down Rattray Street was aggravated when the gripman turned the wheel to apply the screw-operated track brake, in the wrong direction. One passenger (Mr Garret) was killed and several others injured when the tram crashed at the terminus; this unfortunate occurrence apparently did not effect patronage as 800 persons, including some ladies, were reported as having ridden the trams next day.

Nor did such an event appear to impede the subsequent widespread adoption of "pull" curves. Indeed, without a doubt this technological advance by George Duncan quickly rendered cable traction practical for other cities that were not characterised by the straight streets of San Francisco, a circumstance which in turn lead to significant growth in the development of other cable tramway services in North America along with the spread of cable traction to other cities worldwide.

It will be noted from the illustration, that Mr Duncan's "pull" curve was basically a series of cone-shaped pulleys set behind a "chafing" bar which supported the "shank" of the gripper so as to minimise lateral pressure on the gripper jaws. The diameter and frequency of the pulleys depended on the severity and length of the "pull" curve. Over the years modifications by other engineers were to appear in respect to "pull" curve operations but the basic principles first established by Mr Duncan continued to be adhered to.

During construction, these pulleys were required to be set with great precision as was the attendant "chafing" bar, for should a fault develop through failure or misplacement of any "pull" curve component, including the gripman dropping the cable off the pulleys thus letting it get in behind the "chafing" bar, such an event, be it by accident or misadventure, had the potential to create very costly delays and repairs. It has been estimated that at their peak the various cable tramway systems in Dunedin operated about 20 curves of varying radii that made use of the "pull" curve principle.

"Let-go" Curves

On the other hand very few "let-go" curves operated in Dunedin. Two of these occurred on the original single-tracked Rattray Street line where descending grip trams were required to release the cable when negotiating the left-hand passing loops located near York Place and St Joseph's Cathedral. This requirement was eliminated when the line was duplicated in 1884.

For a time a combined "let-go" and "pull" curve also existed on the Roslyn Company's 1900 extension into Kaikorai Valley where a double-tracked loop curved sharply out of Belgrave Crescent into James (now Falcon) Street. Kaikorai Valley-bound grip trams were required to release the rope to traverse the (south side) of this loop before picking it up again in James Street, on the brink of the now single-line, reserved track section located on a I in 3.5 descent. Because it was feared that gripmen might miss "picking up the rope " again at this point, the Public Works Department refused to license this section for passenger operation. A temporary terminus was established at Highgate. Empty-running grip trams en route from the Kaikorai Valley-based car barn to the Highgate terminus each day, however, faced an ascending gradient, and were hauled through the other (northern) side of the loop using the "pull" curve principle.

The Dolphin Brake

In a round about way invention of the "pull" curve and occurrence of the earlier mentioned unfortunate runaway from Cathedral Curve only two months after the original Rattray Street cable line opened for traffic in 1881, may have led to George Smith Duncan's other particular creation: his invention of an emergency brake for cable cars. The device was almost as unique at the cable tram itself.

Described as a "Dolphin" brake, it consisted of a steel wedge fitted to the grip tram that was held directly above "the slot". Operated manually by means of a screw thread, this device which would only be used by the gripman in an emergency, forced a steel wedge down into the slot opening and would effectively halt the car. Any consequential damage which may have arisen was regarded as being preferable to any loss of life.

Although the "dolphin" brake was approved by the Public Works Department by the end of May 1881 and was fitted to all the original grip trams by Cossens & Black, it is not known if the brake was ever subsequently used in an emergency on the Roslyn line. It is known that warping of the wooden "slot" rail ,due to weathering, did cause the wedge to gouge out pieces of the slot rail from time to time; steel slot rails were substituted some time later.

Although the "dolphin" brake was subsequently removed from the original Roslyn grip trams, such a device does not appear to have been fitted to other (later) grip trams in the Roslyn line's fleet nor was it used by any of the other grip trams that operated in Dunedin. It is not known to what extent the "dolphin" brake was adopted by other cable tramways overseas.

However, it is still used today as an essential item of equipment for the cable trams in San Francisco. Here this emergency brake is operated by a lever that forces a steel wedge down into the steel railed "slot". Once applied, the wedge is almost impossible to release; gas torches have to be used to cut the wedge free.

**Mind the Curve! A history of the cable trams*, John D. Keating, 1996

GLOSSARY OF NEW ZEALAND HORSE, STEAM & CABLE-TRAM TERMS

Advertising Boards: Wooden boards placed on the roof, side panels and platform ends.

Aisle: Centre corridor through saloons and open sections between seats.

Apron: The front panel on the end platforms.

Arched Roof: A roof having a continuous symmetrical Curve from one side of the car to the other.

Axlebox: Box shaped casting housing the axle bearing and its associated lubrication system. Transmits the weight of the vehicle to the axle via the springs.

Back Platform: The rear platform of a car on which the conductor stands, as distinguished from the front where the driver stands.

Bearing Plates: On double-truck cars – mounted on the underneath of a car body and on the truck bolster to prevent the body of the car from rocking.

Bell Cord: A cord which rings a bell when pulled. Carried in hangers under the ceiling, extending from end to end of the car to a brass bell on the platform and used by the conductor to give signals to the driver, according to the number of bells rung.

Body: The coach work of a car which sits on the truck or trucks. Includes the underframing, floor, sides and roof, but not the platforms which are usually framed separately.

Body Bolster: The steel beam which supports the swivel double-truck to the body of a tram. The weight of the car body is carried to each truck through a centre bearing mounted on a transverse bolster beam, having side bearing plates to take any outer balance weight. Single-truck cars have no body bolsters.

Bogie: English term for four-wheel swivel truck.

BRAKES:

Hand brake: Operated by the driver or conductor - by a gooseneck handle or wheel mounted on the front platform, having racket wheels at floor level engaging with 'pawls' or 'dogs' which can be kicked into position to hold the brake applied to the wheels.Pressure is applied to the brake shoes on to the wheels by turning the brake wheel or handle on the front platform or back platform.

Brake Shoe: A curved block which is pressed against the tread of a car wheel when the brakes are applied. Brake shoes were made of cast iron in the early years.

Break: A four-wheeled horse-drawn vehicle used in"breaking" horses, had a high seat for the driver and a decoy horse (a trained horse), was placed on the other side of the centre pole to show the new horse what to do.

Bulkhead: The parition extending across the car with a central door leading into the saloon at each end of the car.

Bumper: The front steel horizontal beam to reduce damage in a collision. Made to the contour of the end of the car.

CABLE TRAMWAY TERMS:

Conduit: The tube shaped excavation below the road surface in the centre of the running tracks which housed the moving traction rope. About 2 feet 6 inches deep, the conduit was constructed of concrete and used "yokes" of bent railway iron to support the two running rails and the slot rail in the centre the yokes" alaso provided support for the "carrying pulley" bearings.

Crown pulleys: These were larger than the "carrying pulleys" and were placed at the top of major gradients; about 40 inches in diameter.

Depression pulleys: Sometimes called "kickers" in Dunedin; used at the bottom of major grades to prevent the traction cable from rising out of the slot.

Flying conductor: A conductor who took fares on one cable-car and changed over to another running in the opposite direction.

Gripman: Drove the cable-car by engaging the "gripper" on to the moving rope.

Gripman-conductor: Employed on the Maryhill Extension, Dunedin, to both operate the cable-tram and collect fares.

Gripper: The device that reached down through a slot between the running rails to control the grip-tram's movement. The gripman could take hold or let go the moving cable by the use of this device.

Grip Tram / Dummy (an American term): A vehicle hauled by a mechanically driven endless wire rope housed in an underground tube or conduit.

Grip trap: Two hinged plates which were fitted into the slot rail at certain locations to facilitate the removal of a cable-tram grip should it become damaged.

"Hooking" the rope: A long steel hook which was used to place the rope back into the gripper on short workings like at Highgate on the Roslyn line or when the rope was "lost".

"Let Go" marks: Where gripmen were required to "Let Go" the rope at certain fixed points near the end of a line.

"Pick up" points: Where the track (the roadway surface) dipped so the rope could be picked-up by the gripman by just closing the gripper.

"Pull Curve": A series of drum pulleys in the conduit which allowed the cable-tram to retain hold of the rope while rounding a corner.

Pulleys: The "carrying pulleys" were used to support or guide the moving traction cable through the conduit beneath the roadway; they were about 9 inches in diameter and spaced about 30 feet apart.

Set: A tension control on the gripper which could be altered to allow the gripper to hold the rope more tightly when the cable-tram was heavily loaded.

Sheaves: Large, single grooved wheels that were used to guide the traction cable in and out of the powerhouses. Also used around "let go" and other tight curves and to reverse the cable at terminals.

Shunting: Dunedin Cable-trams were shunted manually in and out of the car-sheds.

Slot: The centre gap between the running rails through which the shank of the grip reached down to the gripping mechanism or "jaws".

Stranding: When one of the six 13-wire strands became worn and began to separate from the other strands. Detection of stranding meant an immediate stoppage of the service to check the condition of the rope. If the stranding was minor, the spot would be marked and the service resumed. After the running of the last cable-tram in the evening, a splicing repair would be done.

Terminal sheaves: These were mounted on a railed trolley called a "tension carriage" that was chained to a counter weight which hung over a 6 feet by 8 feet pit in the powerhouse. This allowed for a constant tension to be maintained on the traction rope at all times regardless of the traffic loading.

The Rope: The term used to refer to the endless steel wire cable which hauled the cable-trams. This consisted of a compound of six strands – each strand containing 13 wires (that is, seven wires over five, over one small flexible wire) using Lang's lay as the rope formation. The six strands were uniformly closed up around the main core of the best quality Russian or European hemp. The breaking strength of the rope, which was 3.5 inches in circumference, was guaranteed as 29 tons.

Trackman: Worker employed for the greasing and oiling of pulleys and sheave wheels; pumping out terminal pits; maintaining the track.

Vault: Any enlargement of the cable conduit to house curve pulleys; terminal sheaves and special equipment.

Canopy: The roof area above the driving platform.

Cant rail: The moulding below the roof on the side of a car. (above the windows).

Change Boys: The name given to the lads who were employed to bring fresh horses in from the depots/stables to the horse-trams at city and suburban terminals and take the tired horses back.

Clydesdale leader horses: This bred of horse was mainly used for working the steep gradients when a heavier horse was required to assist a loaded horse-tram.

Clerestory Roof: A raised section of the roof with an upper row of small windows which gave better ventilation and light in the saloon.

Conductor: A male member of the crew who collects fares from passengers and gives signals to the driver / grip-man to start and stop the car.

Dash: American term for front apron.

Decency Boards: Also known as modesty boards, A panel placed along the upper deck of a tramway vehicle so women in dresses cannot be viewed from a low angle.

Destination signs: Wooden boards mounted on the roof or front panels

Door (sliding): End saloon door which slides into the end bulkhead on the side of the door opening.

Double truck car: A car mounted on two swiveling four-wheel trucks.

Draw Bar: Tramway term for a tow bar. For coupling two cars together.

Drop SashWindow: A window which opens downwards from the top into a pocket in the side of a car below the window sill.

Facing Points: The junction of diverging tracks. A set of points which face the same way as the train is travelling.

Fanlight Window: A small window above a large window. Hinged to open for ventilation of saloon.

Fare Box: Used on one-man cars to collect tickets and coins.

Firebox: Part of a steam tram locomotive boiler where combustion of the fuel takes place.

Flange: Projecting edge or rim on the periphery of a wheel or rail.

Foot-gong: A loud-sounding gong (bell) mounted underneath the platform floor which is sounded by the driver pushing a foot pedal on the platform floor to warn pedestrians of the approach of a car.

Foot-plate: The floor section of a steam-tram locomotive.

Gauge: Distance between the inner edges of the rails of a tram line.

Gradient: Slope or incline to the horizontal of a tramway. Expressed in degrees from the horizontal, as a percentage to the rise and fall or slope length from the horizontal.

Hand straps: Leather slings suspended from a hand strap pole to be grasped by standee passengers standing in the aisle.

Headway: Interval between trams running a timetable service.

Hill boy: The name given to young men and boys employed to ride on the backs of the leader horses up the hills.

Knive-board: Longitudinal seating back-to-back along the centre line on the top-deck of a double-decker, with passengers facing outwards.

Letter board: The wide panel or board on the outside of a car just below the eaves (cant rail). The name of the operator was sometimes painted on this board in the early years.

Lifeguard: Attached underneath the front of a tram to protect a pedestrian or object from being run over. A swinging'gate' at the front of the car which swings backwards on being hit by an object on the road, thereby releasing by a linkage, a steel or wooden tray placed in front of the wheels which drops to the road surface and scoops up the object.

Lift sash: A window sash arranged to raise into a pocket above the windows.

Livery: The distinctive colour scheme of a vehicle.

Monitor Roof: American term for clerestory roof.

Non-smoking: Saloon section reserved for non-smoking passengers.

Panelled sides: Either the flat wooden orsteel panel below the window level of a tram saloon.

Passing loop: A short section of double (two) tracks on a single tram line where trams proceeding in the opposite directions can pass.

Permanent Way: Term used to describe the tram tracks, rails, sleepers, road foundations and the surface around the tracks.

Platform: Generally at a lower level to the passenger accommodation, at either end where the driver of the tram stands and for passenger entry/exit.

Points: Track junction where a car can move from one line to another through moveable tapering rails.

Points drain: Incorporated in the points with a sump to drain away road surface water.

Quarter lights: The end saloon windows in the fixed bulkhead partition at either end of the tram, at right-angles to the side of the saloon. Sliding door between.

Rolling stock: Vehicles used on a tramway.

Running board: A step extending along the side of an open car from bumper to bumper and hung by hangers attached to the body frame.

Saloon: Enclosed passenger accommodation.

SEATS: Generally designed to be durable rather than comfortable. The majority were wooden slatted, some upholstered with heavy carpet material, others with rattan.

Back to back cross bench seat: A continuous seat from one side of the tram to the other with no aisle in the centre, which are placed either side of a body parition – one facing forward, the other to the back of the car. Also known as a bulkhead seat.

Cross bench seat: A continuous wooden seat from one side of the tram to the other with no aisle in the centre.

Garden seat: Transverse reversible seat with a single piece of wood for the back rest.

Longitudinal seat: A continuous seat built along the length of a saloon side which gives more standing room.

Transerve reversible seat: A seat with a back that can be moved backwards and forwards to face the direction the car is moving.

Transverse seat: cross seating at right-angles to the saloon side.

Side Bar: A longitudinal bar placed along open sections of cars to prevent entry or exit at waist level.

Smoking Partition: A saloon partition for passengers who smoke cigarettes or pipes. Usually separated by a bulkhead from the main compartment.

Solebars: The longitudinal structural or chassis members along the sides of the car body frame.

STAIRCASES:

Left-handquarter-turn 90° staircase: A staircase which turns in a spiral to the left - known as direct stairs.

Right-hand quarter-turn 90° staircase: A staircase which turns in a spiral to the right. Known as reversed stairs, the ideal being that a passenger who slipped on the stairs was not likely to be thrown on to the road. Right-handed staircases limited a motorman's view on the left side, and the foot of the stairs was next to the lower saloon entry/exit and became a problem with the flow of passengers.

Steam: Vapour of water resulting from the application of heat. The temperature at which the water boils is dependent upon the pressure to which it is subjected but steam produced remains in saturated form whilst in contact with water from which generated.

Tank locomotive: One which carries its fuel and water supplies on its own main frames.

Tongue and Groove: Boards in which a tongue along one edge fits into a groove along the edge of the next board.

Trailer: non-powered tramway carriage.

Trailing Point: The junction of converging tracks. A set of points which face the opposite way to which the tram is proceeding.

TRAMCAR TYPES

Box car: A small single saloon four-wheel tram. Nicknamed either as 'box cars' because of their small box-like shape.

California Combination Car: The original semi-open car, developed for the California Street Cable Railroad Company, San Francisco. the closed section being in the middle with an open section at either end.

Combination: Combining both closed and open passenger accommodation.Acompromisebetweenclosed andopen cars forchangeble weather conditions.

Double-Decker car: Two decks of passenger accommodation with a staircase at both ends. In New Zealand the upper-deck was open. Slow loading and not being able to use the upper-deck during the winter made this design unpopular with operators.

Open car: A car with no enclosed sides, with tranverse cross bench seats extending the width of the car and full length running boards to permit entry-exit at any point. Canvas blinds provided which can be pulled down between each pair of side posts for passenger protection from wind and rain.

Palace car: Transverse compartment car with facing cross-bench seats in each compartment, sliding side doors for enrty/exit to each compartment from a running board.

Single Saloon: Similar as a box car, although generally a reference to the large double-truck cars with entry/exit platform at both ends of the saloon.

Trammies: Tramwaymen; uniformed men who worked on the horse, steam and cable-trams as inspectors, drivers, gripmen, and conductors.

TRUCKS: The steel frame under a tram which carry the wheels, axles and the braking system.

Single-Truck: Long wheel-base, four-wheel fixed frame truck on which a short car body is mounted. The car body underframe rests directly on the truck springs or on the spring-supported top bar of the truck side frame.

Equal Wheel Truck: Four-wheel swivel trucks on centre bearings of the same wheel diameter. used on long cars in sets of two (eight wheel cars) - one at each end under the saloon.

Valve gear: Mechanism which controls the operation of the steam distribution valve in the steam chest of a steam-tram locomotive cylinder.

Ventilated Arch Roof: An arched roof with in-built ventilators

Underframe: The structure or chassis upon which the car body is built. The body underframe does not include the platform bearers and bumper beams which are usually framed separately and attached to the body underframe.

Waist Panel: The panel immediately below the windows along the side of the saloon.

Wheel base: The measurement / distance between two sets of wheels.

Wheel flange: The raised inner edge of the wheel which projects below the top of the rail and holds the wheels on the rail.

Wrap Round Apron: A curved panel which enclose the front and left-hand side of the driving platform at both ends on rear entrance / exit only cars.

HORSE, STEAM AND CABLE VEHICLES PRESERVED BY MUSEUMS

WANGANUI

FLEET NUMBER	YEAR BUILT	TYPE	DESCRIPTION	MUSEUM	FOR DETAILS REFER PAGE
100	1891	STEAM-TRAM	0-4-0 BALDWIN STEAM-TRAM (R)	MUSEUM OF TRANSPORT & TECHNOLOGY	60

WELLINGTON

FLEET NUMBER	YEAR BUILT	TYPE	DESCRIPTION	MUSEUM	FOR DETAILS REFER PAGE
1	1901	CABLE-TRAM	FOUR-WHEEL CABLE GRIP TRAM (AC)	WELLINGTON CITY COUNCIL	60
2	1901	CABLE-TRAM	FOUR-WHEEL CABLE GRIP TRAM (AC)	WELLINGTON TRAMWAY MUSEUM	60
3	1900s	CABLE-TRAM	FOUR-WHEEL CABLE GRIP TRAM (AC)	IN STORAGE KARORI DEPOT	61
4	UNKNOWN	CABLE-TRAILER	FOUR-WHEEL PALACE (AC)	WELLINGTON CITY COUNCIL	62
5	UNKNOWN	CABLE-TRAILER	FOUR-WHEEL PALACE (AC)	KELBURN PARK, SALAMANCA ROAD	62
6	UNKNOWN	CABLE-TRAILER	FOUR-WHEEL PALACE (AC)	WELLINGTON TRAMWAY MUSEUM	62

CHRISTCHURCH

FLEET NUMBER	YEAR BUILT	TYPE	DESCRIPTION	MUSEUM	FOR DETAILS REFER PAGE
7	1881	STEAM-TRAM	0-4-0 KITSON STEAM-TRAM (R)	TRAMWAY HISTORICAL SOCIETY	82 to 86
10	1894		FORMER HORSE & STEAM TRAM TRAILER, ELECTRIC TRAM TRAILER No. 91. FOUR-WHEEL EIGHT-WINDOW SALOON DOUBLE-DECKER (R)	TRAMWAY HISTORICAL SOCIETY	98, 99, 100, 117, 118, 119
43	1887		FOUR-WHEEL CALIFORNIA COMBINATION; ORIGINALLY FOUR-WHEEL SINGLE SALOON CanTC No. 51; REBUILT AS A CALIFORNIA COMBINATION BY CHRISTCHURCH TRAMWAY COMPANY IN 1897/1898, No. 43; BECAME CTB No. 50 c1912. (R)	TRAMWAY HISTORICAL SOCIETY	92, 93
64	1879		FORMER HORSE & STEAM TRAM TRAILER, ELECTRIC TRAM TRAILER No. 64 FOUR-WHEEL EIGHT-WINDOW SALOON DOUBLE-DECKER (BS)	TRAMWAY HISTORICAL SOCIETY	87, 117, 118, 119
74	1890s		FORMER HORSE & STEAM TRAM TRAILER, ELECTRIC TRAM TRAILER No. 74 FOUR-WHEEL CAGE DOUBLE-DECKER (R)	TRAMWAY HISTORICAL SOCIETY	107, 117, 118, 119
84	1882		FORMER HORSE & STEAM TRAM TRAILER, BUILT ORIGINALLY AS A FOUR-WHEEL SINGLE-DECK SALOON, REBUILT BY CANTERBURY TRAMWAY COMPANY AS DOUBLE-DECKER SEVEN-WINDOW SALOON (BS)	TRAMWAY HISTORICAL SOCIETY	89, 117, 118

DUNEDIN

FLEET NUMBER	YEAR BUILT	TYPE	DESCRIPTION	MUSEUM	FOR DETAILS REFER PAGE
18	1880s	HORSE-TRAM	FOUR-WHEEL SINGLE-SALOON (R)	TRAMWAY HISTORICAL SOCIETY	131, 138
93	1906	CABLE-TRAM	FOUR-WHEEL CABLE GRIP TRAM (BS)	OTAGO SETTLERS MUSEUM	151
95	1905	CABLE-TRAM	FOUR-WHEEL CABLE TRAM (R)	TRAMWAY HISTORICAL SOCIETY	151
103	1903	CABLE-TRAM	FOUR-WHEEL CABLE GRIP TRAM (B)	TRAMWAY HISTORICAL SOCIETY	161
105	1904c	CABLE-TRAM	FOUR-WHEEL CABLE GRIP TRAM (AC)	SEASHORE TROLLEY MUSEUM, KENNEBUNKPORT, MAINE, USA	170
106	1906	CABLE-TRAM	FOUR-WHEEL CABLE GRIP TRAM (ACMG)	OTAGO SETTLERS MUSEUM	172, 173
107	1882	CABLE-TRAM TRAILER	FOUR-WHEEL SINGLE SALOON (R)	MUSEUM OF TRANSPORT & TECHNOLOGY, AUCKLAND	158, 159
110	1880s	CABLE-TRAM TRAILERS	FOUR-WHEEL SINGLE SALOON (BR)	TRAMWAY HISTORICAL SOCIETY	164
111	1920	CABLE-TRAM TRAILER	FOUR-WHEEL SINGLE SALOON (B)	TRAMWAY HISTORICAL SOCIETY	165
125	1900	CABLE-TRAM	FOUR-WHEEL CABLE GRIP TRAM (BS)	OTAGO SETTLERS MUSEUM	175

AC – ACQUIRED IN RUNNING CONDITION
ACMG – ACQUIRED IN RUNNING CONDITION MINUS GRIPPER
BR – BEING RESTORED 1997
BS – BODY ONLY – IN STORAGE
R – RESTORED

Other Grantham House Tramway Books

by Graham Stewart

Always a Tram in Sight

The Electric Trams of New Zealand 1900 to 1964

208 pages – 656 black and white photographs – six pages of colour destinations – same page format as this book – case bound with French folded jacket.

The companion volume to this book – the complete comprehensive book of the New Zealand tram as a public service vehicle. The tramcar had a unique personality, some were elegant and small, others towered like giants over all other traffic. They became familiar friends, part of daily life as they traversed the streets at regualr intervals to a timetable. Tramcar design in New Zealad was a mix of English and American – imports from Britain included classic English double-deckers. Later designs were parochial as local builders were commissioned to build new trams for cities and provincial towns. They became identifiable with the city – the big red monsters of Auckland, the double-saloons of Wellington, the green trams of Christchurch with two trailers snaking behind and the Brill trams of Dunedin.

The End of the Penny Section

First published in 1973 – now revised and enlarged – 272 pages – contains 30 colour plates and 550 halftone photographs – the same page format as this book.

The social history of the tramway age, from the flimsy-looking horse trams on steel rails, the snorting steam tram engines in suburban streets, the magical cable cars that climbed steep hills by invisble power, to the towering electric trams which reigned supreme for half a century, capturing the imagination of all urban New Zealanders. This book not only records the changing modes of passenger transport, but shows the reader the street furniture of each decade, the lamp standards, the style of cars and trucks, the fashions and behaviour of people, the trees on the footpaths, the gardens, the advertising signs. It is full of human interest stories, lots of period humour, and gives an insight into the way families lived in those times.

Around Auckland by Tram in the 1950s

48 pages –- 124 black and white photographs of the city and suburbs – same format page size as this book

A graphic visual statement of Auckland in the middle of this century when Auckland was still a close knit city with the outer boundaries only a tram journey from Queen Street. Everyone took the tram to town – to Queen Street or Karangahape Road, the heart of Auckland in those years – for major household purchases, or for a day out in town, a big event if you were a child.

Motorways and suburban shopping complexes were only at the embryo stage.

Around Dunedin by Cable Car & Tram in the 1950s

48 pages – 75 black and white photographs - large landscape format.

A unique photographic portfolio of Dunedin as it was in the 1950s – the cable cars swarming with commuters, the trams loaded with people and prams, the trammies, the suburbs, the streets, the shops and the advertisements of the period. At each terminus a cluster of small shops – a dairy, a butcher, a grocer, a hairdresser tobacconist, a chemist and a magazine shop.

AUCKLAND
THAMES
PAEROA
WANGANUI
NELSON
WELLINGTON
WESTLAND
CHRISTCHURCH
DUNEDIN
INVERCARGILL